Author Karen Brown
Editor Deneen Celecia
Assistant Editor Linda Milliken
Designer Wendy Loreen
Illustrator Barb Lorseyedi

About the Author

Karen Brown has been a teacher for seventeen years. She has taught learning disabled and elementary students in Lee's Summit, Missouri. Karen has been a finalist in the Lee's Summit Teacher of the Year program. She is a member of the National Education Association, International Reading Association and Society of Children's Book Writers and Illustrators.

METRIC CONVERSION CHART

Refer to this chart when metric conversions are not found within the activity.

4 tsp	=	1 ml	4 cup	=	60 ml	350° F	=	180° C	1 inch	=	2.54 cm
2 tsp	=	2 ml	3 cup	=	80 ml	375° F	=	190° C	1 foot	=	30 cm
1 tsp	=	5 ml	2 cup	=	125 ml	400° F	=	200° C	1 yard	=	91 cm
1 Tbsp	=	15 ml	1 cup	=	250 ml	425° F	=	216° C	1 mile	=	1.6 km

1 oz.	=	28 g
1 lb.	=	.45 kg

www.edupressinc.com

ISBN 1-56472-172-8

Printed in USA

REAL WORLD MATH

Contents

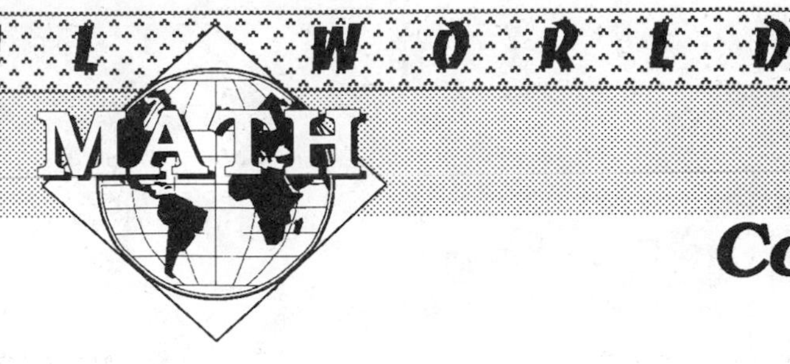

Contents

Content Notes

Exploring the Real World:

Real World Math activities are designed to give students practical knowledge and develop skills for math experiences they will encounter in their everyday lives. No matter what their education level or vocation, these skills are essential to all.

Be sure to review the Table of Contents so that you are familiar with the scope of activities contained in this reproducible book.

About the Pages

The pages are written *to the student* at an upper-grade (4-6) reading level. They adapt themselves well to use in classroom programs, home-schooling situations or learning centers in which the development of independent skills is encouraged (see pages 6-7 for helpful hints and guidelines for center set up and implementation). The activities may also be incorporated by the educator in cooperative group lessons.

Each Real World activity unit is two to four pages in length.

The first page includes:

- ***Your Experience***—Encourages the student to associate the concept introduced with an event in their life.
- ***Let's Explain It***—A short explanation of the real world experience and simple concept introduction.
- ***Practice Page Directions***—Instructions for completing the worksheet/practice page (or pages) that follow.
- ***Words to Know***—Includes a short list of relevant vocabulary.
- ***Things to Think About***—Thought and discussion prompts that encourage critical thinking at all levels.
- ***Real World Activity***—Suggested activities for extending learning into real world experiences.

The second (and sometimes third and fourth) page provides students with the opportunity to practice the skill prior to real world application.

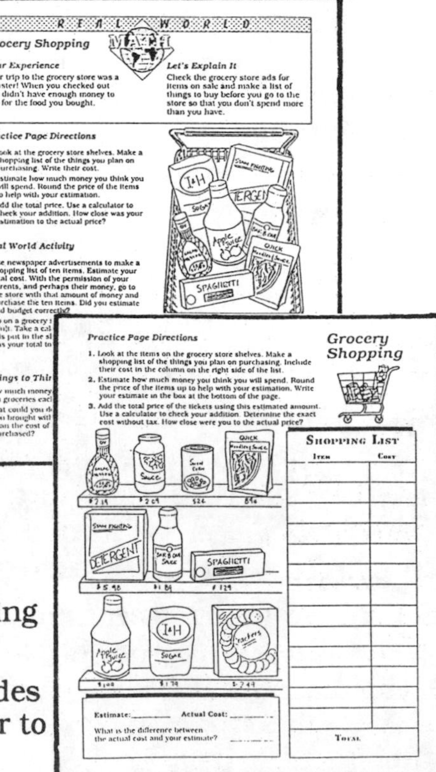

REAL WORLD MATH

Grocery Shopping

Your Experience

Your trip to the grocery store was a disaster! When you checked out you didn't have enough money to pay for the food you bought.

Let's Explain It

Check the grocery store ads for items on sale and make a list of things to buy before you go to the store so that you don't spend more than you have.

Practice Page Directions

1. Look at the grocery store shelves. Make a shopping list of the things you plan on purchasing. Write their cost.
2. Estimate how much money you think you will spend. Round the price of the items to help with your estimation.
3. Add the total price. Use a calculator to check your addition. How close was your estimation to the actual price?

Real World Activity

- Use newspaper advertisements to make a shopping list of ten items. Estimate your total cost. With the permission of your parents, and perhaps their money, go to the store with that amount of money and purchase the ten items. Did you estimate and budget correctly?

Practice Page Directions

Grocery Shopping

1. Look at the items on the grocery store shelves. Make a shopping list of the things you plan on purchasing. Include their cost in the column on the right side of the list.
2. Estimate how much money you think you will spend. Round the price of the items up to help with your estimation. Write your estimate in the box at the bottom of the page.
3. Add the total price of the tickets using this estimated amount. Use a calculator to check your addition. Determine the exact cost without tax. How close were you to the actual price?

Estimate: ________ Actual Cost: ________

What is the difference between the actual cost and your estimate? ________

SHOPPING LIST	
ITEM	COST
TOTAL	

Real World Math Centers

The real world activities may be used in learning centers if you prefer. The activity pages are written at student level to facilitate independent or cooperative learning.

Set Up Tips

- Introduce no more than five activity units at a time. When students have demonstrated understanding and the enthusiasm for more information, introduce more units. Be sure you have selected a variety of interest areas.
- Laminate (optional) and put each card in a box with the necessary supplies. Use a marker to make a supply list on the side of the box.
- Reproduce practice pages and place in the box.

Preparing Students

Review the sections of each activity card.

- Encourage students to talk about their own experiences and the thought stimulators in the **Things to Think About** section. Comment on the importance of reading the explanation at the top of each page.
- Point out the supply list and ask them to keep track of the materials in each activity box. Have them inform you when supplies are missing or low.
- Discuss the purpose of the practice pages and the location of their directions.
- Foster the use of the related words in their conversations about the activities.
- Explain the *Real World Journal* (page 6). Work together to prepare a notebook for their journal entries. Give examples of things they might include.

Tracking Progress

Here are some suggestions for keeping track of activities completed:

- Make individual contracts. Allow students to select areas of interest to pursue and explore.
- Make a chart to post on the wall. Award a star for each task card completed.
- Use the Table of Contents as an individual contract. Photocopy one for each student participating and ask them to highlight or put a sticker next to each unit completed.

Follow-up

- Set aside discussion and sharing time to evaluate and share the activities pursued in and out of the classroom. Allow them to share the contents of their Real World Journals.
- If you are a teacher, encourage parents to point out and pursue the connections between school curriculum and real world experiences. If you're home schooling, focus on activities that integrate daily life with specific reading concepts.

Real World Journal

Parent, Teacher Information

Keeping a journal is a great way for children to express their opinions, develop a specific interest and practice organizational skills. It's also a great way for parents and teachers to assess skill development and encourage areas of interest.

A *Real World Journal* is similar to a scrapbook about everyday life and the practical skills required to succeed in it! Encourage your child or your students to update and add information to their journal as they make new discoveries about mathematics and how it relates to their everyday experiences. As they get involved in the real world activities suggested in this book, there are many references to gathering information about a topic and keeping a record of it in their *Real World Journal*. The words always appear in *italic* lettering so be on the lookout for these opportunities. As students get more involved, they will begin to find and add information to their *Real World Journal* on their own.

An old three-ring notebook can easily be decorated and gain new life as a *Real World Journal*. Provide children with a box of supplies to have on hand to make adding articles and information a simple and pleasurable pastime. Include:

- stapler
- glue
- clear tape
- lined paper
- paper clips
- lined notebook paper
- hole punch
- scissors

Photocopy a stack of the *Real World Journal* page that follows. It can be hole-punched and used as decorative paper for recording opinions, illustrations and observations.

Always remember that the *Real World Journal* is a personal scrapbook that is not intended to be graded or judged!

37.45
7
8
9
÷
4
5
6
X
3
2
3
-
0
.
=
+
ON
MC
MR
M+

REAL WORLD MATH

Credit Cards

Your Experience

Your parents hand a plastic card to the cashier and say, "Charge it!" Does this plastic card take the place of money? Do you have to pay for things you charge?

Let's Explain It

Your parents will eventually pay for items charged (purchased). The credit card company charges interest on any items not paid in full by the end of the billing period.

Practice Page Directions

1. Sign the credit card. Decide which items you wish to purchase.
2. Write down the cost of each item purchased. Calculate the answers to fill in the blanks on the statement.

Real World Activity

- Ask your parents if they would share with you one of their credit card statements.
- Find out what the interest rate is.
- For what purposes did they use the card?
- How much is the minimum payment?

Things to Think About

Why would national credit cards be able to charge less interest than department store cards?

When and why would using a credit card become a problem?

What interest would you pay if you only paid the minimum payment for a year?

Words to Know

- interest
- account
- finance charges
- credit
- statement
- charge

1. Sign the credit card.
2. Circle the items you want to charge.
3. Write the name of each circled item and its cost on the credit card statement.
4. Fill in all the blankcs on the statement. Calculate minimum payment and interest charges.

Student Bank VISA

555 444 333 222

Expiration date 5/05
Name:

Student Bank / VISA Statement

- Account Number 555 444 333 222
- Balance from previous month: $0.00
- Amount of new purchases: __________ *(Add everything you wrote on the statement.)*
- Minimum amount due: (10%) __________ *(Multiply 10% times the amount you spent.)*
- Credit Limit $5,000.00
- Credit Available: __________ *(Subtract what you spent from the credit limit.)*

Your Name:

Your Address:

Purchases

Item	Cost

Annual Percentage Rage 18%

If not paid in full, a finance charge of _____ % will be added to next month's bill.

(Multiply 18% times your total charges for this month.)

REAL WORLD MATH

Classified Advertisements

Your Experience

Sometimes when people are looking for something to buy or sell they check the newspaper.

Let's Explain It

The classified section in the newspaper is where people can look for or advertise things for sale.

Practice Page Directions

1. Pull out the classified section from the newspaper.
2. Find an ad that mentions a purchase price or money for each category.
3. Tape it underneath the category name.

Real World Activity

- Call your local newspaper to find out the rates for placing a classified advertisement for a garage sale. How much do they charge per word? Do you need to have a minimum amount of words to place an ad?

Things to Think About

Is the advertising cost reasonably priced compared to the profit you will make?

How many people will receive your ad if you place one?

Is it better to advertise in a newspaper that is a local publication?

Words to Know

- advertisement
- issues
- publication
- profit margin
- classified
- miscellaneous

Classified Advertisements

Practice Page Directions

1. Find an ad for each category from the classified section of the newspaper.
2. Tape it underneath the category name and circle the part of the ad that mentions money.

Help Wanted/Job Offered

Auto For Sale

Apartment For Rent

Horse For Sale

House For Sale

Furniture For Sale

Boat For Sale

Dog or Cat For Sale

REAL WORLD MATH

Checkbook

Your Experience

You see people paying with paper that they have written on and signed instead of paying with money.

Let's Explain It

A check is the same as money. You have to open a checking account at a bank and have money in the account for at least the amount of the check.

Practice Page Directions

1. Complete the ledger that has the information already there for you. **Remember:** *Add* a deposit and *subtract* a check. Ask someone to help you with the first one.
2. Cut apart the checks on the following page. You may want to make more copies before you do that. Compare the check with one from someone's checkbook. Find out what all the information on the check means.
3. Practice writing some checks.
4. Complete the blank ledger using the checks you wrote. The beginning balance is already there for you.

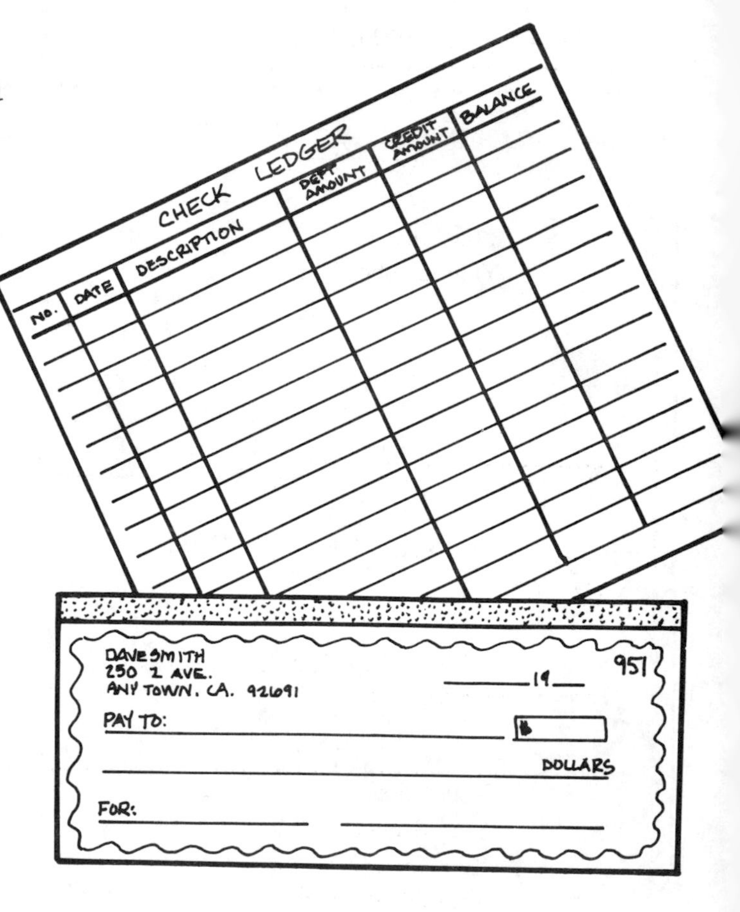

Real World Activity

- Visit a bank near your home. Ask for a blank check ledger book. They usually have some to give away. Talk to the tellers and managers. Be ready with three questions you want to ask them.

Things to Think About

What happens when the money in your account is less than what you wrote the check for?

Why were banks opened in the first place?

How do you open a checking account?

Words to Know

- balance
- checkbook
- entry
- ledger
- account
- deposit
- overdrawn
- bank

Checkbook

Your Name
Your Address
Your City, State, Zip Code
Your telephone number

Student Bank
3333 Main Street
Your City
70-7978/1377

Date ____________________

Pay to the order of ____ $ __________

____ DOLLARS

MEMO ______________________________
What is the check for?

____
Your signature

0046 721 380 009222 •••503240

Your Name
Your Address
Your City, State, Zip Code
Your telephone number

Student Bank
3333 Main Street
Your City
70-7978/1377

Date ____________________

Pay to the order of ____ $ __________

____ DOLLARS

MEMO ______________________________
What is the check for?

____
Your signature

0046 721 380 009222 •••503240

Your Name
Your Address
Your City, State, Zip Code
Your telephone number

Student Bank
3333 Main Street
Your City
70-7978/1377

Date ____________________

Pay to the order of ____ $ __________

____ DOLLARS

MEMO ______________________________
What is the check for?

____
Your signature

0046 721 380 009222 •••503240

Checkbook *Checking Account Ledger*

Number	Date	Description of check	Payment/ Debit —	Deposit/ Credit +	Balance $826.39
		Toy World	$15.22		
		birthday gift for Sue			
		Corner Market	$53.55		
		groceries			
		The Bike Shop	$22.79		
		bike repair			
		Deposit		$115.55	
		Sports World	$37.79		
		soccer shoes			
		Deposit		$725.65	
		Pet Company	$6.95		
		dog food & bones			
		Dr. Smith	$125.00		
		checkup			
		Main Street Drug Store	$45.56		
		prescription			
		Clothing Plus	$16.38		
		socks, t-shirts			

Checkbook

Checking Account Ledger

Number	Date	Description of check	Payment/ Debit —	Deposit/ Credit +	Balance $725.00

Buying Produce

Your Experience

When you are in the produce section of the market, people put fruits and vegetables on scales.

Let's Explain It

Fruits and vegetables are often priced by the pound. When people want to know how much they are spending on the produce they buy, they need to weigh it first.

Practice Page Directions

1. Look at the produce in each scale. Figure out how many pounds it weighs.
2. Look at the price per pound. Figure out how much it would cost to buy the produce in the scale. Show how you got the answer. Did you add or multiply?

Real World Activity

- Go with your mom or dad to the market. Help them weigh the fruits or vegetables they buy. Read the scale to see if you can figure out the total weight.
- Find the sign that tells the price per pound. Figure out what the cost is going to be. Write it down in your *Real World Journal* or notebook. Check your answers against the register receipt.

Things to Think About

Why are some fruits and vegetables not sold by their weight?

Why do the prices for produce change?

Words to Know

- per pound
- multiply
- ounces
- scale
- weight
- produce

Practice Page Directions

Buying Produce

1. Look at the produce in each scale. Figure out how many pounds it weighs.
2. Look at the price per pound. Figure out how much it would cost to buy the produce in the scale. Show how you got the answer. Did you add or multiply?

Peaches
79¢ per pound

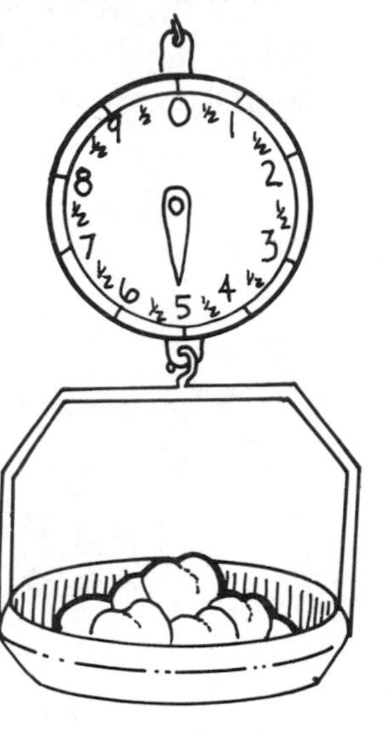

Work here:

Answer:

Zucchini
63¢ per pound

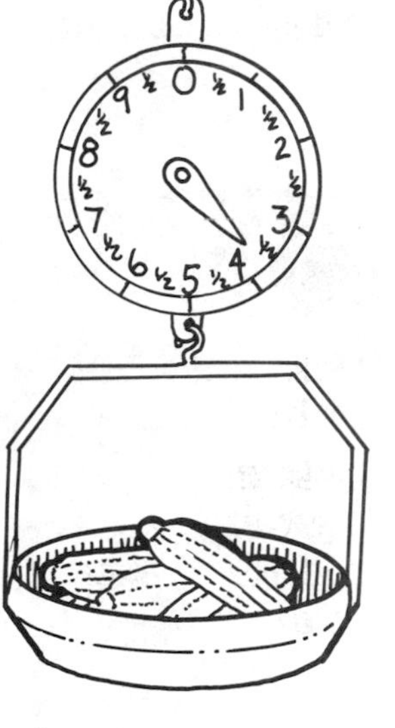

Work here:

Answer:

Oranges
75¢ per pound

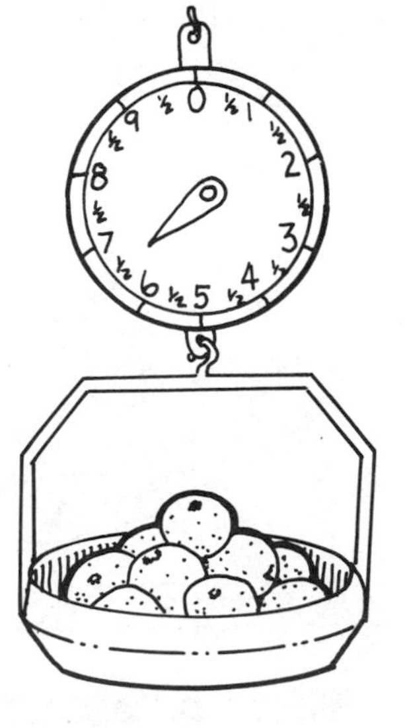

Work here:

Answer:

Broccoli
1.02¢ per pound

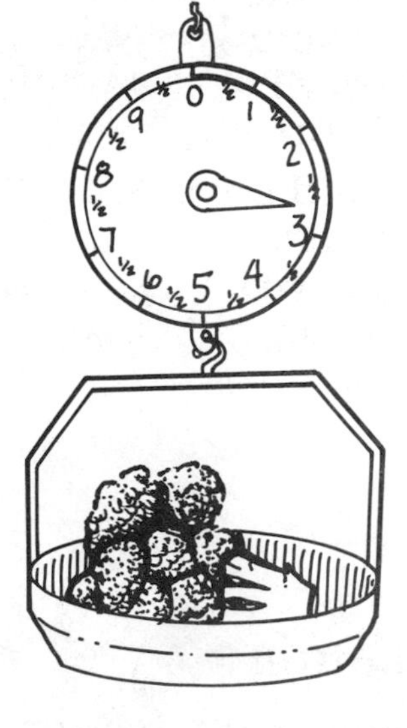

Work here:

Answer:

REAL WORLD MATH

Sports Scores

Your Experience

You like to play a lot of sports but each one scores points differently. Sometimes that gets confusing!

Let's Explain It

Each sport is different and has its own set of rules for play and scoring. Scoring determines the winner of an event.

Practice Page Directions

1. Look at the point values for each game.
2. Calculate how many points each team scored.
3. Decide the final score and circle the winning team on the scorecard.

Real World Activity

- Attend a football, baseball, soccer, hockey, tennis or basketball game. Keep track of the points scored in your *Real World Journal.*
- Sit next to the scorekeeper if one is in the audience.
- Go bowling with friends and learn how to keep track of the points scored.

Things to Think About

How does scoring vary from sport to sport?

Do the best players usually win?

Are the games scored in the same manner in other countries?

Words to Know

- game
- scorekeeper
- tournament
- champion
- game
- winner

Practice Page Directions

Sport Scores

1. Look at the Football game and the Basketball game statistics.
2. Calculate the scores according to the points assigned for the achievement.
3. Decide the final score for each team and circle the winner.

Football

Touchdown—6 pts.

Field Goal—3 pts.

Point after—1 pt.

Safety—2 pts.

Conversion—2 pts.

TEAM A	Points	TEAM B	Points
1st Quarter 1 touchdown 1 field goal		**1st Quarter** 1 field goal	
2nd Quarter 1 safety		**2nd Quarter** 1 touchdown 1 extra point	
3rd Quarter 2 touchdowns 1 point after		**3rd Quarter** 1 touchdown 1 safety	
4th Quarter 1 field goal		**4th Quarter** 1 touchdown 1 conversion	
Total		Total	

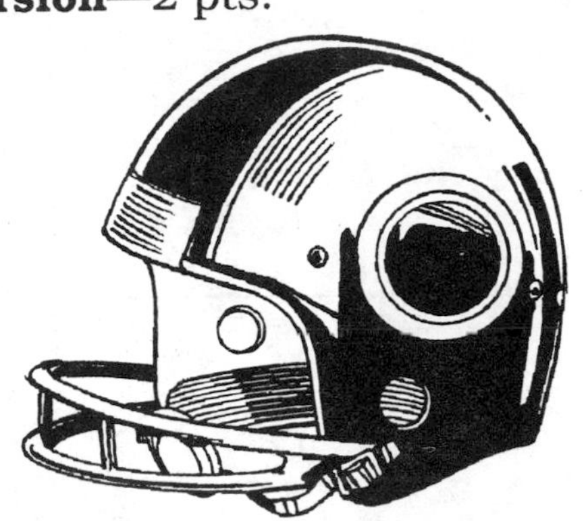

Basket—2 pts.
Free Throw—1 pt.
Outside of Range—3 pts.

Basketball

TEAM A	POINTS	TEAM B	POINTS
1st Quarter 10 baskets 2 free throws		**1st Quarter** 15 baskets 1 free throw	
2nd Quarter 6 baskets 2 free throws 2 baskets in 3 pt. range		**2nd Quarter** 10 baskets 3 free throws 4 baskets in 3 pt. range	
3rd Quarter 8 baskets 3 free throws 3 baskets in 3 pt. range		**3rd Quarter** 6 baskets 1 free throw 2 baskets in 3 pt. range	
4th Quarter 11 baskets 4 free throws		**4th Quarter** 7 baskets 2 free throws	
Total		Total	

REAL WORLD MATH

Sports Statistics

Your Experience

When listening to a sports broadcast you hear the announcer mention "batting average and RBI". Why do they keep track of these statistics?

Let's Explain It

Statistics are calculated numbers that help show strengths and weaknesses in players and teams so that improvements and game plans can be made.

Practice Page Directions

1. Study the Sandlot's statistics for the season. Calculate the batting average of each player.
2. Rewrite the batting order according to their performances, from highest to lowest.
3. If you were the coach, who would you give the awards to at the end of the season? Study the team's statistics first!

Real World Activity

- If you play on a team, ask your coach to show you the statistics that are kept on your team. If you are not on a team, locate a team whose coach will talk to you.
- Find out how these statistics are used.
- Compare your team's statistics with those of a professional ball player or team.

Things to Think About

How are statistics different from sport to sport?

Are the best players with the best statistics usually the most valuable on the team?

Are the games scored in the same manner in other countries?

Words to Know

- statistics
- performance
- R.B.I.
- averages
- statistician
- percentage

Sport Statistics

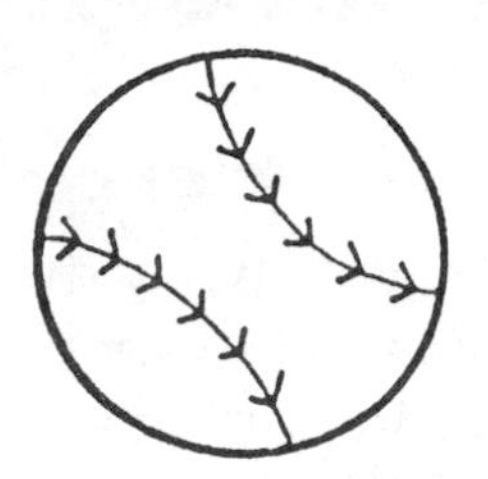

Practice Page Directions

1. Study the Sandlot's statistics for the season. Calculate the batting average of each player. Remember: batting average equals the number of hits divided by the number of times at bat.
2. Rewrite the batting order according to their batting averages, from the highest to the lowest.
3. If you were the coach, who would you give the awards to at the end of the season? Study the team's statistics first!

THE SANDLOTS **Season Statistics**

	#Hits	# at bat	R.B.I.	1B	2B	3B	H.R.	Bat. Avg.	Errors
Brett	7	42	3	3	2	1	1		6
Otis	4	40	1	2	1	1	0		4
Arnold	8	16	2	4	1	2	1		3
Samantha	8	25	1	8	1	1	3		3
Frank	5	35	2	4	0	0	1		0
Jose	6	48	3	3	2	1	0		1
Thomas	9	40	3	6	2	0	4		2
Pat	3	28	2	0	0	1	0		3
Martin	12	50	0	2	0	0	0		4

R.B.I.= Runs Batted In H.R.= Home Runs 1B= Singles 2B= Doubles 3B= Triples

Batting Order

1.

2.

3.

4.

5.

6.

7.

8.

9.

Awards

Home Run King

Top Hitter

Least Errors

Most Valuable

REAL WORLD MATH

Calories & Grams

Your Experience

You hear many adults say they are on a diet. They are counting calories or fat grams. What is so important about these two things?

Let's Explain It

A calorie is a measurable unit of energy supplied by a food. Fat grams tell how much of a food is made up of fat.

Practice Page Directions

1. Look at the side panel of the box of cereal.
2. Answer the questions and show your work.
3. Look at the page from the calorie and fat counting book.
4. Calculate the calories and fat grams eaten that day.

Real World Activity

- Write down all that you eat in one day in your *Real World Journal.*
- Look up the calorie and fat grams in each food item in a book or on the package.
- Add up the amounts to determine how many calories and fat grams you ate during the day.

Things to Think About

How many calories or fat grams should be eaten in one day?

Does body weight determine how many calories or fat grams to be consumed?

How many calories are used up in different activities?

Words to Know

- calorie
- nutrients
- fat
- energy
- diet
- grams

Calories and Grams

Practice Page Directions

Look at the cereal box panel and the page from the calorie and fat counting book and answer the questions.

1. How many calories would you be eating if you ate two cups of cereal? How many calories would be from fat?

2. How many fat grams would you consume if you ate the whole box of cereal? Protein?

3. Is this cereal or your favorite cereal better for you? Why? Look at the nutritional information on a box of your favorite cereal. What other comparisons can you make?

Nutrition Facts
Serving Size 1 cup
Servings per container 12

Amount per serving
Calories 130
Calories from fat 5
Total fat 0.5g
Total Carbohydrates 30g
Protein 2g

	Fat Grams	Calories
Apples	0	81
Bananas	0	109
Cereal	1.5	89
Donuts	7.4	156
Eggs	5.3	78
Fish	5.6	126
Gumdrops	0.2	98
Hamburger	15.7	231
Lemonade	0	99
Milk with chocolate	8.5	208

4. Calculate the calories and fat grams in the following meal:
   3 eggs, 1 donut, a glass of milk

5. What would be a balanced breakfast with fewer calories and fat grams?

6. Calculate the fat grams in 25 gumdrops.

7. How many calories are in two glasses of lemonade? What price would you charge for each glass at a lemonade stand if you charged per calorie? How much would you earn if you sold ten glasses?

8. How many more grams of carbohydrates than grams of fat are there in one cup of cereal?

REAL WORLD MATH

Making Change

Your Experience

When you purchase something at the store and you hand them more money than it costs, the cashier gives you change back. How can you be sure it is correct?

Let's Explain It

The clerk will usually count back your change to you by starting with the price the item cost and counting up to the amount you've given.

Practice Page Directions

1. Look at the items for sale. Each has a price tag.
2. Total the amount you would spend if you bought the items listed in each box.
3. Calculate the amount of change you would receive if you paid with a $10.00 bill. Next calculate change if you gave a $20.00 bill.

Real World Activity

- Ask your parents to allow you to do the purchasing of all items paid for in cash for one week.
- Take a calculator with you to the store to see if your total is the same as the cash register's. Don't forget to find out what the tax rate is for your city and add it on.
- Decide what change you should receive from the cashier and see if it's what is handed back.

Things to Think About

Why should you make change using the least amount of coins possible?

What it the purpose of taxes? How is the tax rate determined?

Could you calculate your change in another country?

Words to Know

- sales tax
- cash register
- amount
- cashier
- change
- money

Practice Page Directions

Making Change

1. Look at the items for sale. Each has a price tag.
2. Total the amount you would spend if you bought the items listed in each box.
3. Calculate the amount of change you would receive if you paid with a $10.00 bill. Next calculate change if you gave a $20.00 bill.

baseball, mitt, dice, ice cream cone

Total spent =

$10.00 – ______ Change =

$20.00 – ______ Change =

baseball, marbles, doll, teddy bear

Total spent =

$10.00 – ______ Change =

$20.00 – ______ Change =

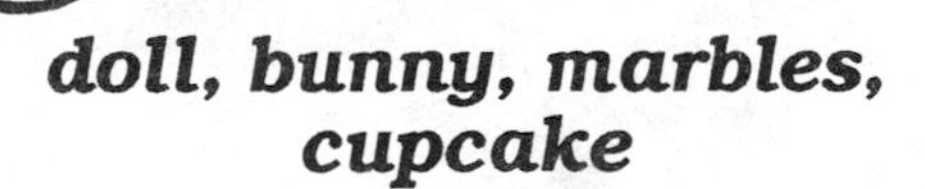

doll, bunny, marbles, cupcake

Total spent =

$10.00 – ______ Change =

$20.00 – ______ Change =

teddy bear, doll, marbles, baseball

Total spent =

$10.00 – ______ Change =

$20.00 – ______ Change =

REAL WORLD MATH

Food Fractions

Your Experience

You always have to share the food with other family members. How can each person get an equal amount?

Let's Explain It

It is easiest to count how many people will be sharing the food, so that equal proportions for each person can be divided.

Practice Page Directions

1. Look at the foods on the next page. You must cut them into equal-sized portions to share your food with your friends.
2. Experiment in pencil to divide up the food until it is in equal pieces.

Real World Activity

- Visit a pizza parlor, cookie factory or donut shop. Watch how they divide the portions to make them equal.
- Take a miniature package of candy pieces. Divide it equally with your best friends.
- Ask a waiter or waitress to share with you how a restaurant cuts pies and cakes into equal-sized portions.

Things to Think About

How are large, medium and small pizzas divided differently?

What is a baker's dozen?

What kinds of things can be divided into equal portions by counting?

Words to Know

- portions
- divide
- proportions
- equal
- fractions
- divide

Practice Page Directions

Food Fractions

1. Look at the foods on the page. You must divide the food into equal portions to share with four friends.
2. Use a pencil to divide each of the foods into equal portions.
3. Practice on paper then practice with food!

REAL WORLD MATH

Coupons

Your Experience

You watch your parents cut and save coupons for food and other items then take them to the store when they shop.

Let's Explain It

When a coupon is given to the cashier at the market, money is subtracted from the bill.

Practice Page Directions

1. *Page 29*—Read the coupons at the top of the page and answer the questions.
2. *Page 30*—Choose ten coupons from the food section of your newspaper. Tape them inside the empty box. Use the coupons to answer the questions below the box.
3. *Page 31*—Make a grocery list for your family. Find coupons for as many items as you can. Answer the questions below.

Real World Activity

- Visit a local grocery or drug store. Ask who is responsible for mailing in the coupons received at their store. Do they return these daily, weekly or monthly?
- Cut out three coupons from a magazine or newspaper. Take them with you the next time someone in your family goes shopping. Calculate how much money was saved.

Things to Think About

Do any stores in your area offer double coupons?

Who pays the doubled amount?

How much money do you think your family would save in a month by using coupons?

Words to Know

- coupon
- manufacturer
- double coupon
- redeem
- expiration
- value

Practice Page Directions

Coupons

Answer the questions below about the coupons.

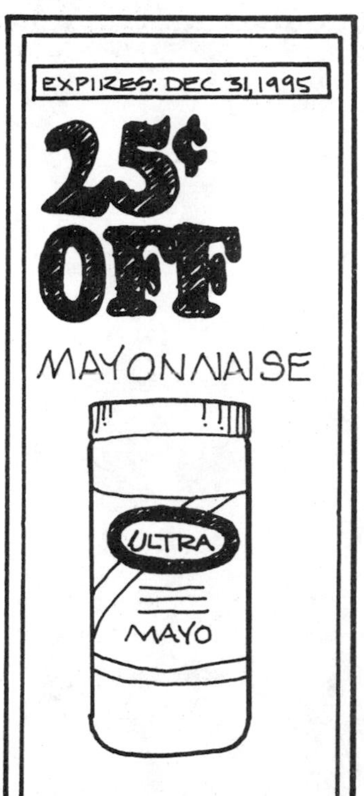

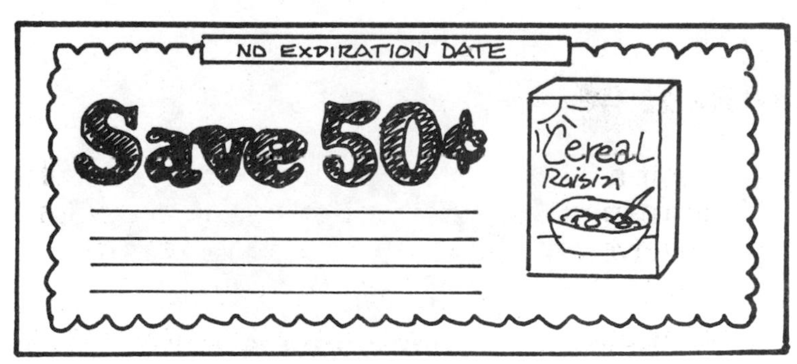

1. What is the total value of all the coupons?

2. What is the **average** discount on these coupons?

3. Circle the expiration date on each coupon. How many have NO expiration date?

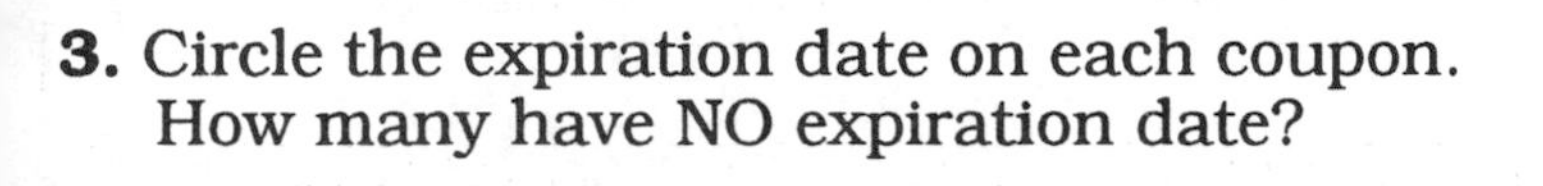

4. How much would you save if you bought toothpaste on a double-coupon day?

5. Which coupon offers the biggest savings?

6. How much would you pay for a jar of mayonnaise if it cost $1.50 and you used the coupon above?

Coupons

Choose and cut ten coupons from the food section of the newspaper. Tape or glue them in the box.

1. What is the total value of all the coupons?

How much would you save if you doubled the savings?

2. Check the expiration dates on the coupons. What is the earliest date?

What is the latest date?

3. If you could use all these coupons three times, how much would you save?

4. Which coupons offer the biggest savings?

5. What is the **average** discount on these coupons?

Coupons

Make a grocery list for your family. Find coupons for as many of the items as you can. Answer the questions below.

Grocery List

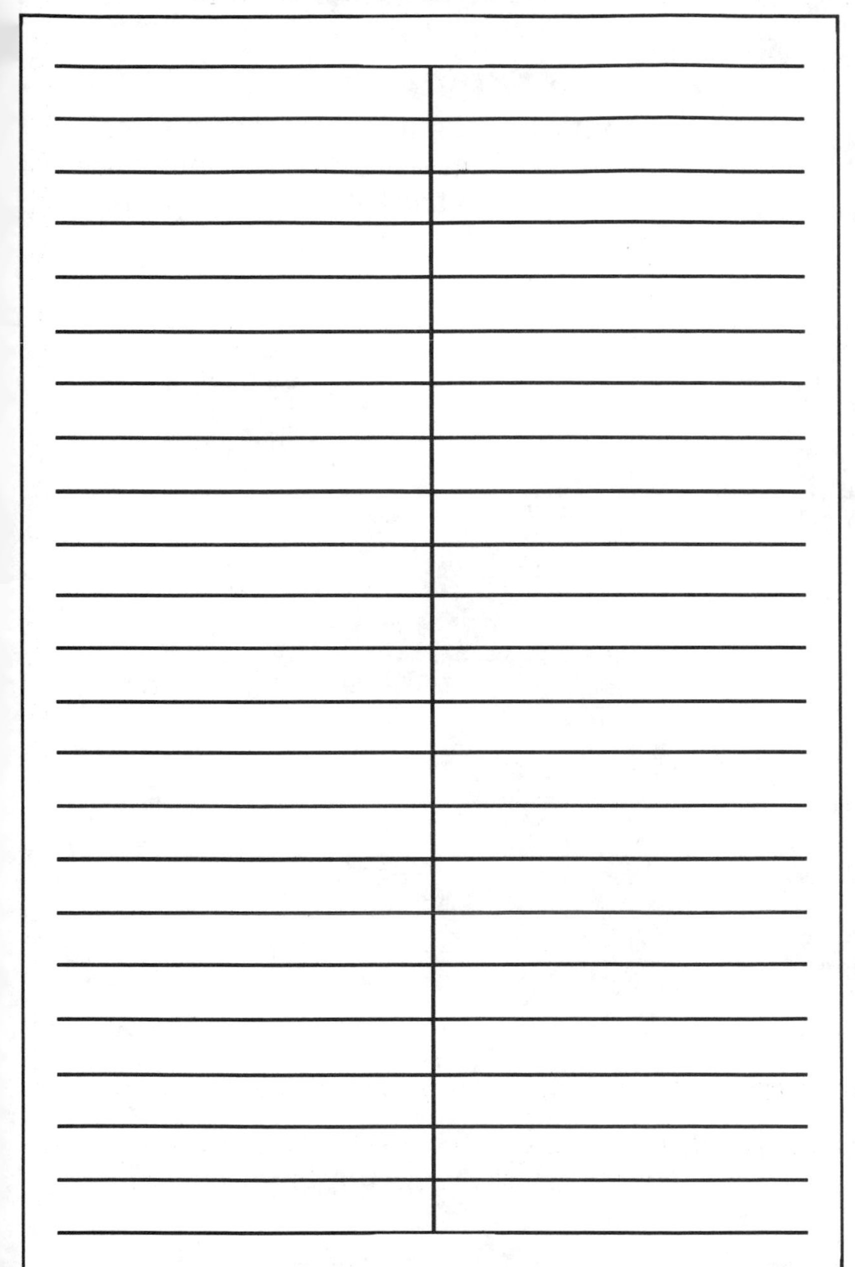

1. How much money would you be able to save by using coupons? Add the coupons here.

2. How much would you save if the coupons were doubled?

REAL WORLD MATH

Budgets

Your Experience

You've often heard your parents and maybe even business people or politicians say, "It's not in the budget." What do they mean?

Let's Explain It

Many homes and businesses make up a budget—a plan for spending money—so that money is spent wisely and appropriately.

Practice Page Directions

1. Look at the amount of money that you have to spend. Fill in a budgeted amount for each category. There is room to add three more categories.
2. When you are done, add all the numbers in each category. If you have "gone over" (budgeted more than you have to spend), go back and change the budget until you have it "balanced".

Real World Activity

- Use graph or ledger paper (available in office supply stores) to set up a budget based on your allowance. Keep track of how much you earn and how much you have to spend in each category.
- Set a budget then plan a short trip or outing with your family. Keep track of your expenses. Compare the end results with your budget plan.
- Look through the newspaper and cut out the articles that are about budgets. Paste them in your *Real World Journal.*

Things to Think About

What happens if you spend more money than you have in your budget?

What items or budget categories could you do without in your budget?

What happens when a government spends more than they have budgeted?

Words to Know

- deficit
- expenses
- balanced
- budget
- income
- categories

Practice Page Directions

Budgets

1. Look at the amount of money that you have to spend. Fill in a budgeted amount for each category. There is room to add three more categories.
2. When you are done, add all the numbers in each category. If you have "gone over" (budgeted more than you have to spend), go back and change the budget until you have it "balanced".

Total Budget

$450.00

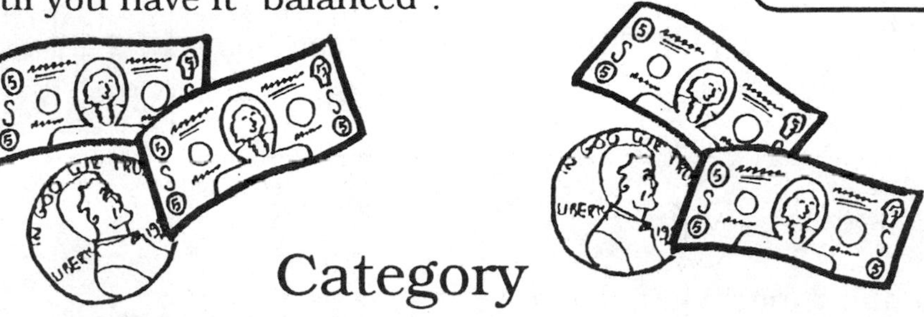

Work Space

Category	Budgeted Amount
Food	
Rent	
Clothing	
Medical	
School Supplies	
Telephone Bill	

Budget total

REAL WORLD MATH

Allowance

Your Experience

Your parents have agreed to pay you an allowance for chores you do around the house.

Let's Explain It

Getting an allowance is a way to earn money before you are old enough to have a job. If you work hard and save wisely you'll have money to pay for things you want.

Practice Page Directions

1. Look at the weekly allowances across the top of the chart. Calculate how many weeks you would have to save your allowance in order to purchase the item alongside the chart.
2. Complete the chart by writing your answer in the box next to each item to purchase.

Real World Activity

- Come up with a plan to earn an allowance. If you already earn one, create a plan that will enable you to earn more. Decide on the chores or responsibilities you will carry out. Discuss with your parents the amount of money you would be able to earn.
- Be responsible for keeping track of the chores you complete and the allowance you earn.

Things to Think About

Is the allowance you receive approximately the same as other friends your age?

What items are you responsible for buying with your allowance?

Should children get allowances for chores they perform?

Words to Know

- allowance
- income
- responsibilities
- budget
- savings
- percentage

Practice Page Directions

Allowance

1. Look at the weekly allowances across the top of the chart. Calculate how many weeks you would have to save your allowance in order to purchase the item alongside the chart. Use fractions of weeks, if necessary.
2. Complete the chart by writing your answer in the box next to each item to purchase.

Allowance per week	$5.00	$8.00	$3.50	$2.50	$7.00
 Bike $98.00	**20 Weeks**				
 Doll $18.00					
 Disney World Ticket $26.00					
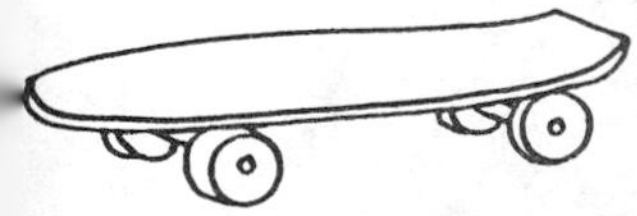 Skateboard $42.00					
 Book $10.00					

REAL WORLD MATH

Postage Stamps

Your Experience

A letter you mailed to a friend you met at camp was returned to you with a message "Returned—no postage."

Let's Explain It

A letter or package must have a stamp on it in order to be mailed. Its cost depends on its weight. Stamps must be authorized by the postal service.

Practice Page Directions

1. Look at the weight of each package or letter. Look at the postage chart to help you calculate the amount of stamps needed to mail it.
2. Put an X through the stamps you would need to purchase.

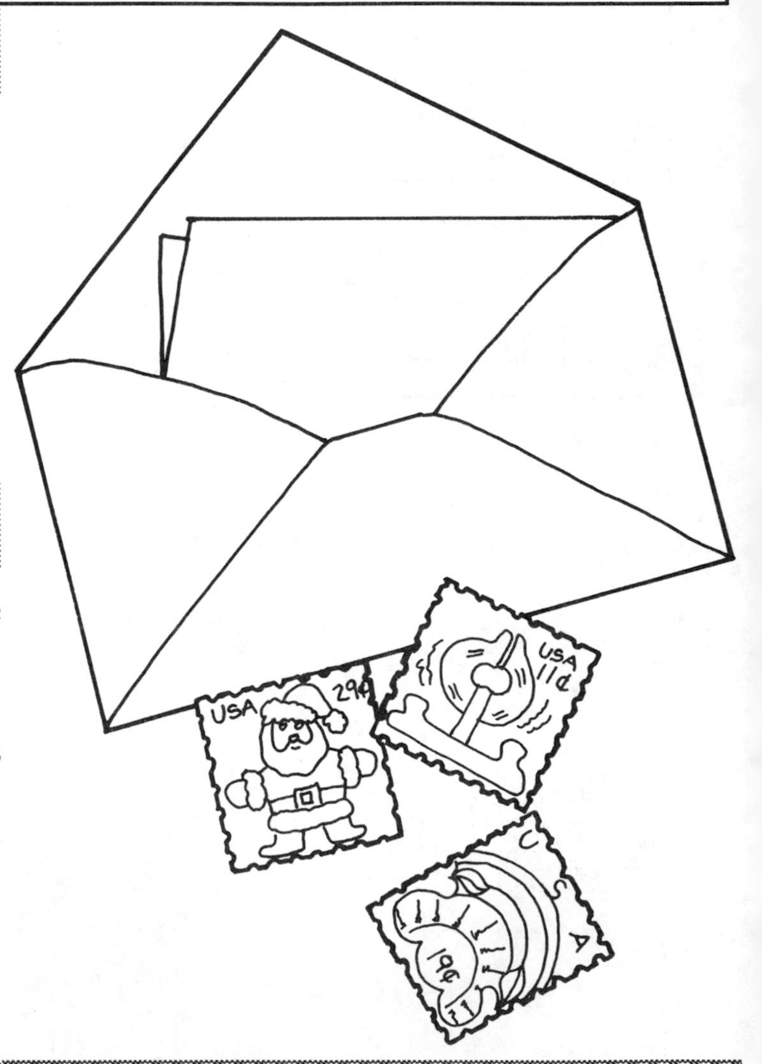

Real World Activity

- Write a letter or mail a package to someone you know. Weigh it and determine the amount it would cost. Purchase the stamps. Find out the difference in price if you were to mail it first class or third class (parcel post).
- Ask your post office if tours are ever given.

Things to Think About

Why must you tell what the contents of a package are when mailing it to another country?

Is the current rate for postage stamps and postcards reasonable?

Why are postcards less expensive to mail?

Words to Know

- parcel
- rural route
- zip code
- air mail
- postmaster
- P.O. Box

Postage Stamps

Practice Page Directions

1. Look at the weight of each letter. Look at the postage chart to help you calculate the amount of stamps needed to mail it.
2. Put an X through the stamp (or stamps) under each letter you would need to purchase.

Postage Rates

Letters	
Up to one ounce	$.29
Each additional ounce	.23
Post cards	.19
International letters	.50
Up to 1/2 ounce	

1 ounce

Carrie Becker
967 Harper Court
Smith, KS 00954

4 ounces

Anita Parker
5 Orange Street
New York, NY 02111

post card

Jason Smith
22 Main Street
Phoenix, AZ 98795

2 ounces

Leanne Milliken
78 Setter Dr.
Point, CA 99902

1/2 ounce

Maria Lopez
Route 2
San Ramon SPAIN

REAL WORLD MATH

Movie Outing

Your Experience

For entertainment, your family and friends enjoy seeing movies at the theater. There are so many choices and decisions to make.

Let's Explain It

An outing can be expensive! With planning, there are ways to be conservative and economical with your money and still enjoy the activities you like.

Practice Page Directions

1. Use the charts to plan an outing to the movie theater.
2. Calculate the costs in the cheapest way, then the most expensive.
3. Compare the difference in cost.
4. Decide which one is best for you and your friends.

Real World Activity

- Plan a trip to go to the movies with your friends or family.
- Calculate the costs ahead of time and show the figures to your friends and parents. Decide if you can afford it. Who will pay for it? How can you save money?

Things to Think About

Do theaters allow you to bring your own snacks into a movie? Why or why not?

How much does it cost the theater to show the movie?

What effect does age have on ticket prices?

Words to Know

- matinee
- savings
- concession
- economical
- budget
- tickets

Movie Outing

Practice Page Directions

1. Plan a movie outing for a **family of four—2 adults, one student and one child under 12** years of age.
2. Calculate the costs according to the instructions in each box below.
3. Compare the difference in cost.

Times	Ticket Prices	Concessions	
Evening: After 5:00 p.m.	$2.75 under 12 $3.25 Students $6.50 Adults	Popcorn	$2.25
		Licorice	1.00
		Drinks:	
		Small	1.25
		Medium	1.50
		Large	1.75
Matinee Before 5:00 p.m.	$3.00 All ages	Candy	1.00
		Gum	.50
		Nachos	2.50

Matinee Nachos for everyone 4 medium drinks **Total Cost:**	Evening movie 2 popcorn 1-licorice 3 small drinks 1-bag candy **Total Cost:**
Evening movie 4 popcorn 4-gum 4 large drinks 4-bag candy **Total Cost:**	Matinee 2 licorice 2 bag candy 4 small drinks **Total Cost:**

- Put an X through the outing that would be most expensive.
- Circle the one that would be least expensive.

R E A L W O R L D MATH

Income

Your Experience

Your parents work in order to purchase the items your family needs and wants. You would like to have a job for extra income, too.

Let's Explain It

People must earn an income—money—to pay for living expenses. Some run their own business, some work for a company and bring home a paycheck.

Practice Page Directions

1. *Page 41*—Complete the invoice and total the charges for the services provided.
2. *Page 42*—Make a flyer to hand out to possible customers.
3. *Page 43*—Complete the information on the invoice. Make several copies to give to customers when you have completed a job for them.

Real World Activity

- Ask people who are in business for themselves how they got started.
- Visit an advertising agency to see the benefit of different forms of advertising.
- Take a survey of adults that you know to determine what jobs they would pay a young adult to do for them.
- Start a business. Use the advertising flyer and invoice on the following pages.

Things to Think About

What will you do if you get too much business? Too little?

Would your own parents get a discount on your services?

What jobs are best charged by the hour rather than by the service?

Words to Know

- income/wages
- discount
- paycheck
- invoice/bill
- fee/charges
- service

Practice Page Directions

Job Income

1. A business must give a customer an *invoice* in order to be paid for a completed job.
2. Complete the invoice below. The service charges listed in the box will give you the information you need.

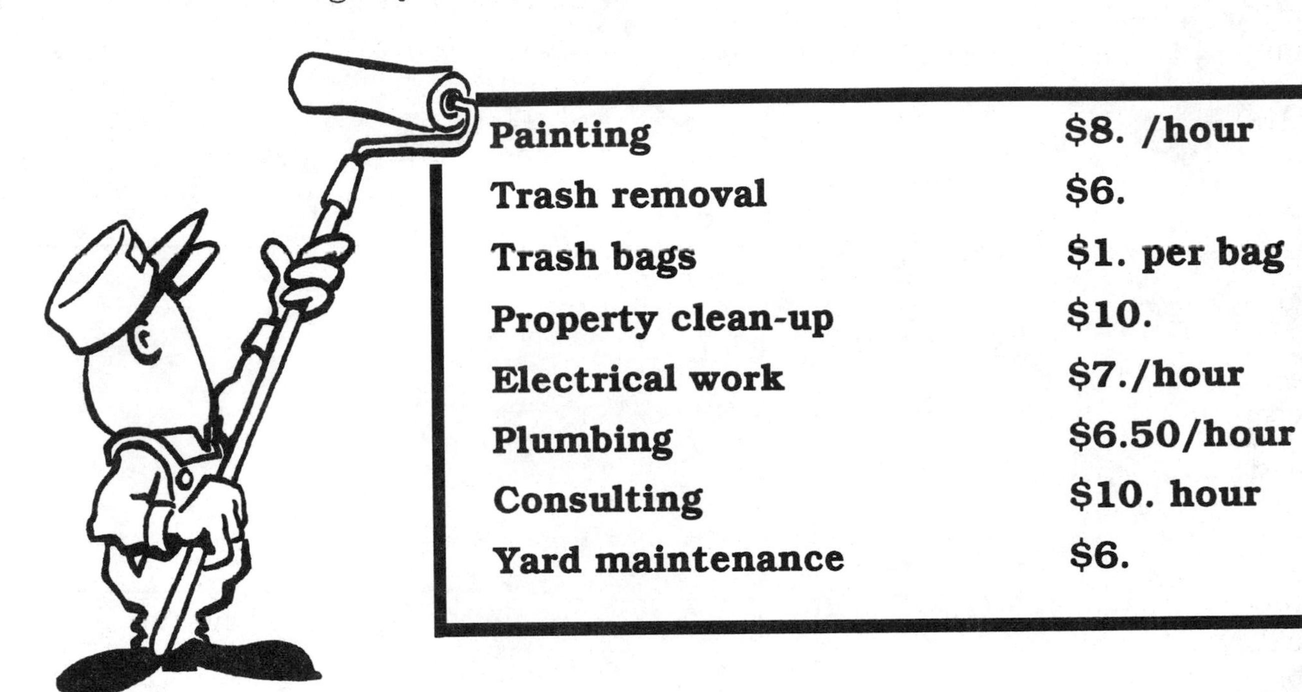

Painting	**$8. /hour**
Trash removal	**$6.**
Trash bags	**$1. per bag**
Property clean-up	**$10.**
Electrical work	**$7./hour**
Plumbing	**$6.50/hour**
Consulting	**$10. hour**
Yard maintenance	**$6.**

Job/Service Description	# Hours	Hourly Rate	Charges
Electrical wiring	2		
Consulting	3		
Painting	12		
Trash Removal			
2 Trash bags			
		Invoice total	

Practice Page Directions

1. Think about what you are good at doing. List all jobs you are capable of doing for a household or business.
2. Consider the value to others and to yourself to determine a reasonable price. You can charge by the job or by the hour.
3. Complete a final draft of the flyer. Make copies to give to neighbors and businesses to advertise your services.

Handy Helper Services

These are the jobs I can help you with and what I charge for my services:

Job	Charges

References:

These are the people who would recommend me.

Call me TODAY

at phone #

Invoice

Handy Helper Services

Date

Customer Name

Address

City

Business Phone#

Please make payment to:

Job/Service Description	# Hours	Hourly Rate	Charges
		Invoice total	

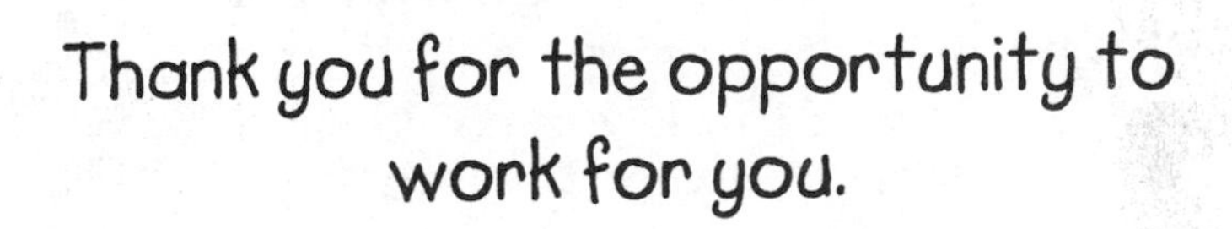

Sale Prices

Your Experience

You want to get the most out of your hard-earned money! You've been told to look for things "on sale" in the stores.

Let's Explain It

Stores often offer discounts on products to encourage customers to buy. The discounts are as much as 50% or 25%. This means you save money off the original price!

Practice Page Directions

1. Go shopping during City Department Store's grand opening sale. Figure out how much you will pay for each item.
2. Calculate how much you saved.

Real World Activity

- Go to the mall with some money you saved for just that purpose. Don't buy anything unless it is on sale. Keep track of the original price, the sale price and how much money you saved.
- Check the receipts to make sure the discounts were given to you.

Things to Think About

Why do stores have sales during different times of the year?

How much do you think you could save over the course of a year if you bought all your clothes at a sale price?

Words to Know

- percent
- sale
- one third-33%
- discount
- savings
- half-50%

Practice Page Directions

Sale Prices

1. Go shopping during City Department Store's grand opening sale. Figure out how much you will pay for each item.
2. Calculate how much you saved.

This Rack—25% off

Everything on this table half price!

10% off selected items

Choose two items to buy from each sale table or rack. Write them below. Write the original price and the sale price. Figure your savings. Calculate your **total** savings.

Item	Original Price	Sale Price	Savings
		Total Savings	

REAL WORLD MATH

Fund-raising

Your Experience

The students in your school are selling gift wrap to earn money for school activities. How much money does your school get from every item sold?

Let's Explain It

Some companies give a percentage of the profits—a certain amount for every item sold—back to the organization that is selling their product.

Practice Page Directions

1. Main Street School just had a fund-raiser. The students had a variety of things to sell. The results are shown.
2. Calculate the total amount of money raised *(gross profit)*.
3. Figure the amount the school gets to keep for school activities *(net profit)*.
4. Calculate how much the school raised all together.

Real World Activity

- If you belong to a club or youth group that sells products to earn money, ask to be involved in figuring total sales and profits.
- Look through magazines for ways you or the club you belong to could earn money. Write for information. Decide on the most profitable product to sell. Record the information in your *Real World Journal.*

Things to Think About

What things have people come to your house selling door-to-door? Is this a good way to earn a living?

How do groups raise money for charity if they do not sell certain products?

Why are some items more profitable?

Words to Know

- profit
- fund-raising
- net profit
- percentage
- charity
- gross profit

Practice Page Directions

Fund-raising

1. Main Street School just had a fund-raiser. The students had a variety of things to sell. The results are shown.
2. Calculate the total amount of money raised *(gross profit)*.
3. Figure the amount the school gets to keep for school activities *(net profit)*.
4. Calculate how much the school raised all together.

Gift Wrap

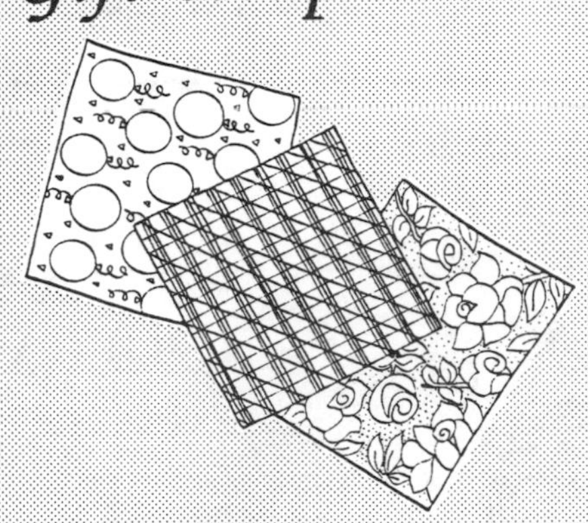

Total Sold: 125 rolls **Cost per roll:** $2.25

Gross Profit: $

School share: $1.25 per roll

Net profit School: **Company:**

Candy Bars

Total Sold: 217 bars **Cost per bar:** $1.00

Gross Profit: $

School share: $ .35 per bar

Net profit School: **Company:**

Calendars

Total Sold: 106 calendars **Cost per calendar:** $3.00

Gross Profit: $

School share: $1.45 per calendar

Net profit School: **Company:**

What was the total amount the school raised?

What do you think they could afford to spend it on?

REAL WORLD MATH

Clocks

Your Experience
There are several types of clocks in your home. The electricity has gone out and you need to reset them. You know how important it is that they are all accurate.

Let's Explain It
There are two types of clocks you may find in your home. Digital clocks display the time in numerals. Standard clocks use hands that point to numbers.

Practice Page Directions

1. Read the number words for time under each pair of clocks. Rewrite the time on the two different types of clocks.
   - Draw the hands in the standard clock.
   - Write the numerals in the digital clock.

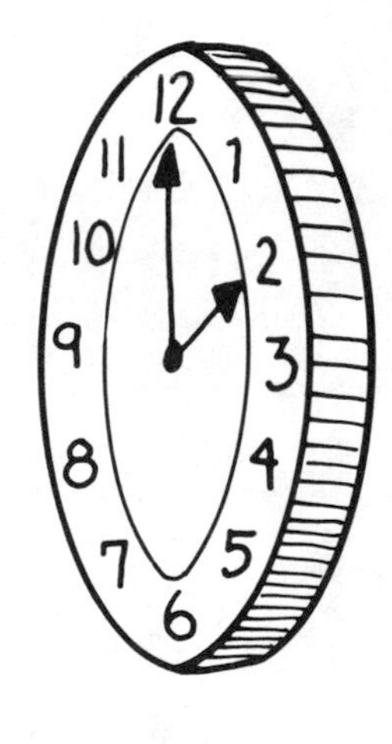

Real World Activity

- Be the official "clock keeper" in your home. Make it your job to set all the clocks whenever there is a power failure, a change in daylight savings or when someone needs an alarm to be set.
- Visit a person who makes or repairs clocks, if there is one in the area.

Things to Think About
Who invented the digital clock?

What is the purpose of daylight savings time? When was it started?

Can you name the parts of a standard clock. How do each of the parts work?

How can an alarm on a clock be helpful?

Words to Know
- digital
- standard clock
- alarm clock
- increments
- hands
- clock face

Practice Page Directions

Clocks

1. Read the number words for time under each pair of clocks. Rewrite the time on the two different types of clocks.
   - Draw the hands in the standard clock.
   - Write the numerals in the digital clock.

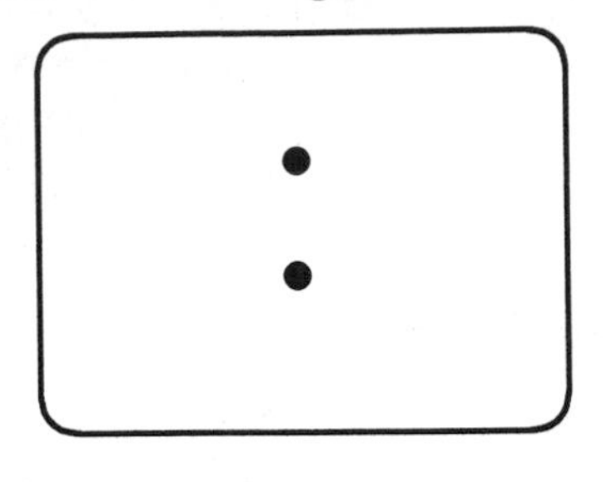

1. five minutes after nine

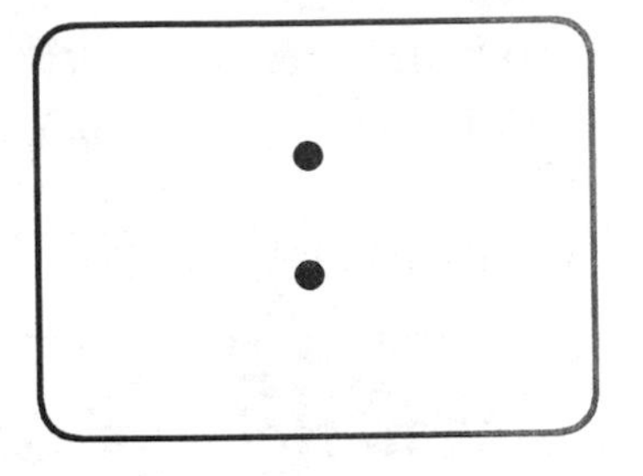

2. forty-five minutes after six

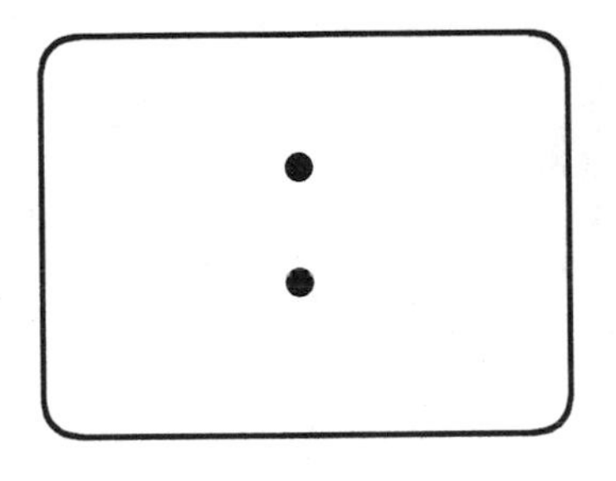

3. half past one

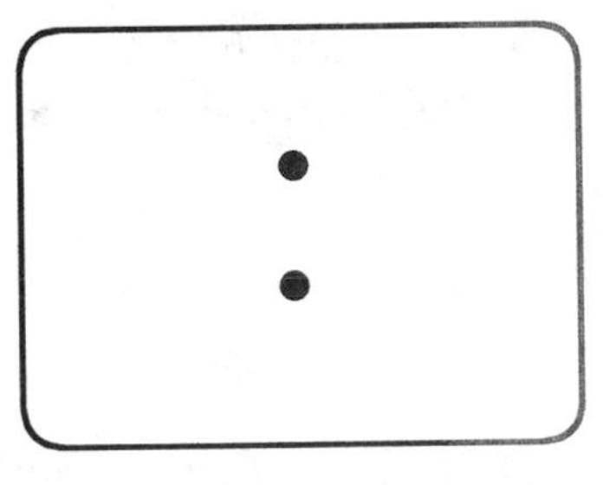

4. quarter past four

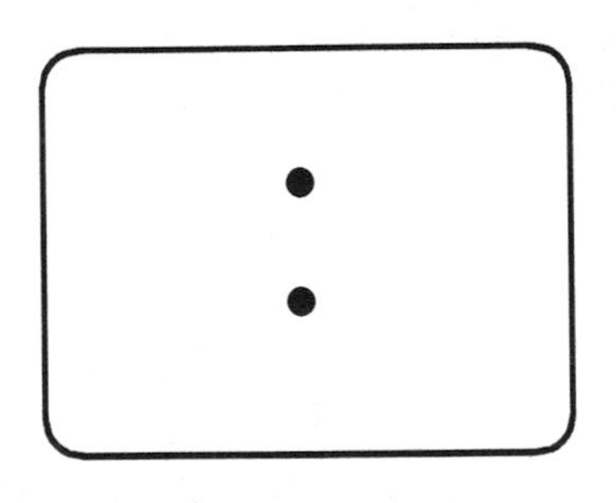

5. midnight

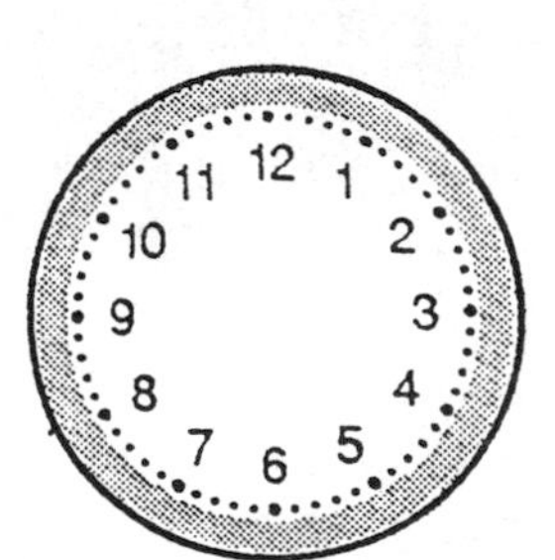
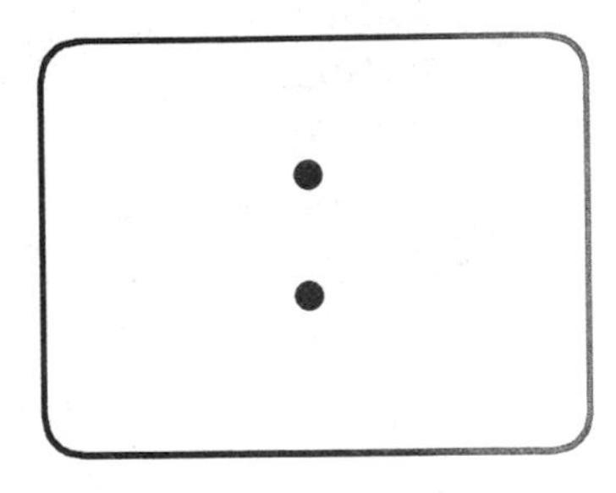

6. seven twenty-five

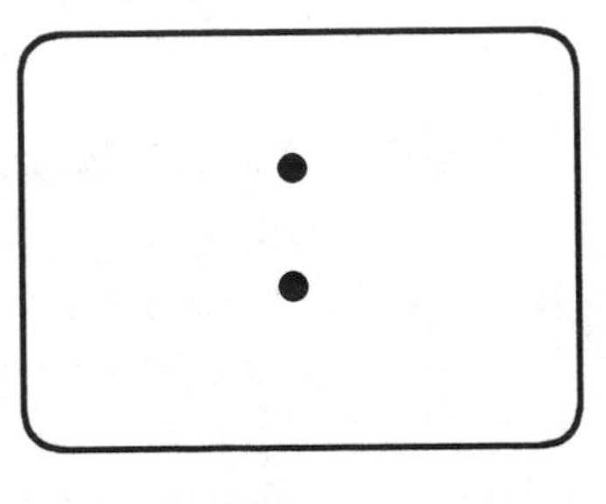

7. ten minutes until four

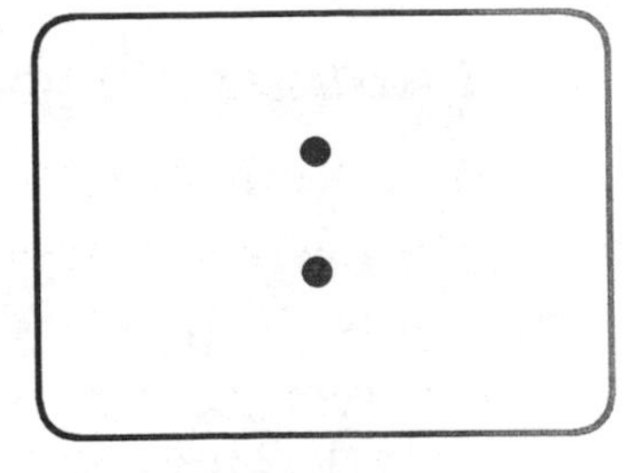

8. ten minutes after eight

REAL WORLD MATH

Automobile Gauges

Your Experience

There's a variety of gauges that the driver looks at while driving an automobile. What are they for?

Let's Explain It

Each car has a panel with gauges that gives the driver different information about the car's performance and condition.

Practice Page Directions

1. Look at the dashboards on each of the cars on the following page. Answer the questions in the column to the right.
2. Add up the number of correct answers to see if you're ready to drive.

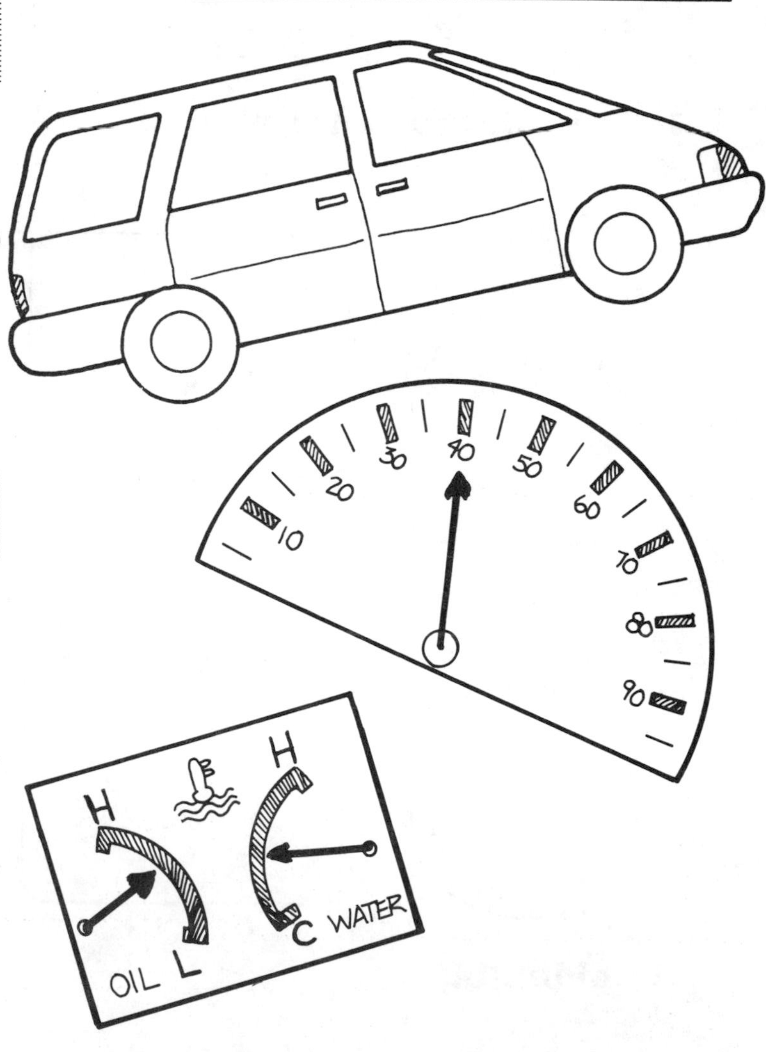

Real World Activity

- Look at the odometer in your family car(s).
- Record the total number of miles driven on your car.
- Guess how far one mile (or kilometer) is (2 or 5 or 10). Have your parents tell you how close you were to being correct.

Things to Think About

Why are all these gauges needed?

Could one gauge be eliminated?

Think of an instrument that would be valuable to add to a car.

What would metric gauges look like?

Words to Know

- tachometer
- speedometer
- odometer
- mileage
- gauge
- speed limit

Practice Page Directions

1. Look at the dashboards on each of the cars. Answer the questions.
2. Add up the number of correct answers to see if you're ready to drive.

2 points= stay off the road
5 points= practice a little more
8 points= ready for your license

Automobile Gauges

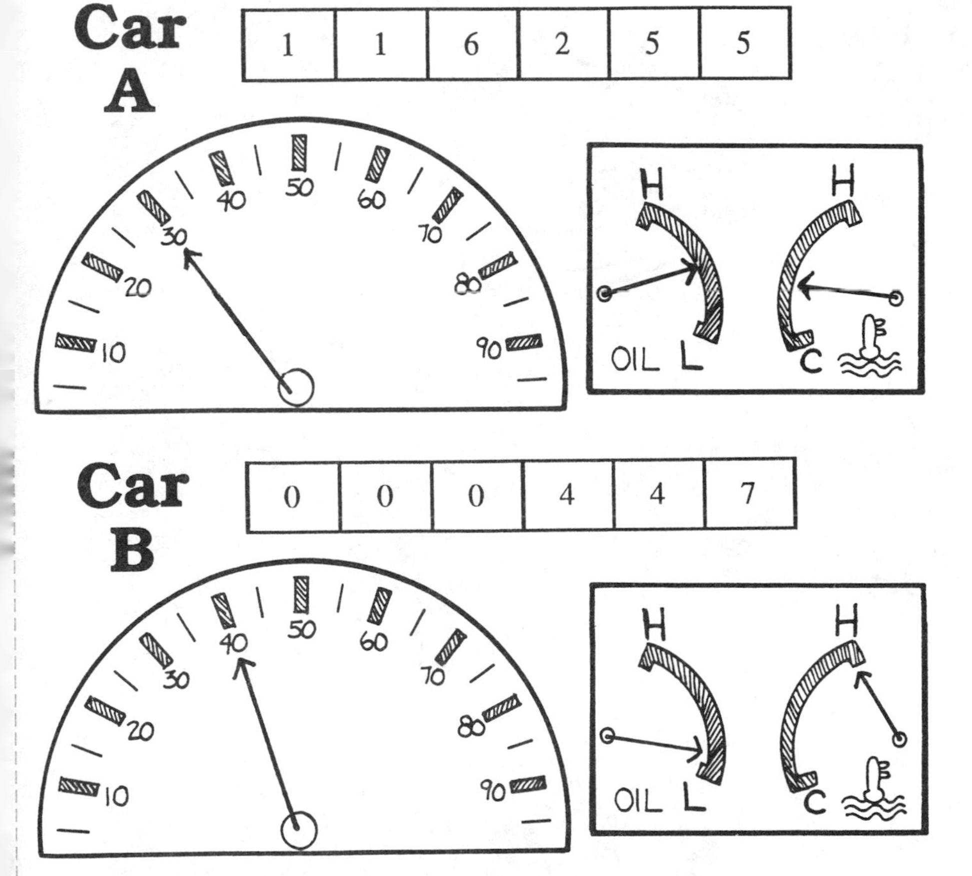

Car C

0	6	5	9	9	8

10 20 30 40 50 60 70 80 90

H H

OIL L C

Write the car's letter in the space below each question.

1. Which car has the least oil?
2. Which car is going faster than 55 miles per hour?
3. Which car is overheating?
4. Which car has been driven the least?
5. Which car has more than 100,000 miles on it?

Calculate the answers

6. How many more miles does car A have on it than Car B?
7. How much faster is car C going than car B?
8. If you need a tune-up every 5,000 miles, how many tune-ups has car A had?

Answers:
1. B 2. C 3. B 4. B 5. A
6. 115808 miles 7. 25 miles per hour
8. 23 tune-ups

R E A L   W O R L D

MATH

Buying a Car

Your Experience

Your older brother is shopping for a car. He is trying to find a low interest rate. Why is that important?

Let's Explain It

Many people pay for expensive things over a period of time. If you borrow money to do this, the lender charges interest on the money you borrow.

Practice Page Directions

1. Look at the cars from which your brother has to choose.
2. Figure their total cost. Find out how to do this on the practice page.
3. Figure the monthly payment and penalty fees. Fill out the payment coupon under each car.

Real World Activity

- Ask if your parents or other relatives have a car payment. What was the cost of the car? What interest rate did they pay? How long will their payments last?
- Visit a car lot. Explain to the salesperson that you are working on a school project. Look at the "sticker prices" in the car windows. Ask what the interest rate and payments would be on a four-year loan.

Things to Think About

What other expensive items are there?

What happens if you cannot make a payment?

What is a down-payment?

What interest rate does your bank give on savings accounts?

Words to Know

- down payment
- borrow
- payments
- loan
- interest
- balance

Practice Page Directions

Buying A Car

1. Look at the cars from which your brother has to choose.
2. Multiply the price by the interest rate, then add that number to the original price to figure the total cost.
3. Figure the monthly payment. To do this, divide the total cost by the number of **months** of the length of the loan.
4. Complete the monthly payment coupon under each car. Figure the late penalty by multiplying the payment by 5%.

Price of Car: $7,500
Interest Rate: 10%
Length of Loan: 3 years

PAYMENT COUPON

Monthly payment:

5% late payment penalty:

Price of Car: $12,500
Interest Rate: 10%
Length of Loan: 4 years

PAYMENT COUPON

Monthly payment:

5% late payment penalty:

Price of Car: $17,350
Interest Rate: 8%
Length of Loan: 5 years

PAYMENT COUPON

Monthly payment:

5% late payment penalty:

Price of Car: $23,000
Interest Rate: 9%
Length of Loan: 5 years

PAYMENT COUPON

Monthly payment:

5% late payment penalty:

Calculators

Your Experience

You've been handed a small instrument to calculate your math problems. What is this instrument and how does it work?

Let's Explain It

The tool is called a *calculator*. It figures the answers to all types of math equations. It has a memory to store numbers if necessary when figuring.

Practice Page Directions

1. Work the math problems by hand. Time yourself.
2. Now use a calculator to check your answers again. Time yourself.
3. Subtract one time from the other to find the difference in how quickly you can perform with a calculator and without.

Real World Activity

- Ask your teacher if you can use a calculator on certain problems in the classroom.
- Use a calculator to check your parents' subtraction in their checkbook. They will appreciate it if you find any errors that they may have made.

Things to Think About

What makes the calculator so accurate?

How is a cash register like a calculator?

What uses does the calculator have in business?

Words to Know

- calculator
- equations
- product
- addend
- dividend
- divisor

Practice Page Directions

1. Work the math problems below by hand, timing yourself.
2. Now use a calculator to check your answers again. Time yourself.
3. Subtract one time from the other to find the difference in how quickly you can perform with a calculator and without.

25 x 42=	**160 -:- 4=**	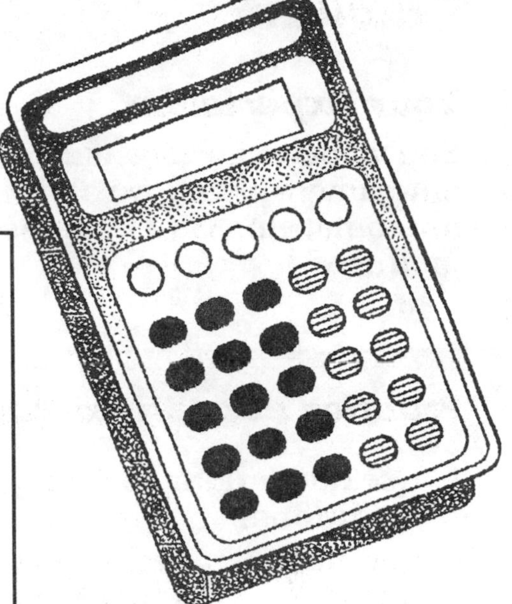
2580 + 3719=	**8691 - 3176=**	**2 x 18 x 4 + 200**
(477 -:- 30) - 12=	**679 - 214 + 75 -:- 6=**	**90 + (13 x 4)=**
5948 - 4679=	**(150 -:- 3) x 4=**	

R E A L W O R L D MATH

Calendar

Your Experience

You've been given a calendar as a gift. How will you be able to use it for something other than super pictures?

Let's Explain It

Calendars tell us which day of the year it is. They can help with planning and organizing time.

Practice Page Directions

1. Use a calendar to help you answer the questions.

JUNE

SUN	MON	TUE	WED	THUR	FRI	SAT
	1	2	3	4	5	6
7	8	9	10	11	12	13
14	15	16	17	18	19	20
21	22	23	24	25	26	27
28	29	30	31			

Real World Activity

- Use a calendar to mark all important dates from homework assignments to birthdays to holidays. Figure out, from time to time, how many days, weeks or months there are until the actual event.
- Visit a first or second grade classroom that practices on the calendar each day as a group. What math skills do these children use when doing it?

Things to Think About

How and when was a calendar first utilized?

Has the calendar changed any over the years?

What is leap year?

Words to Know

- month
- calendar
- leap year
- annual
- yearly
- ordinal

Practice Page Directions

Calendar Math

Use a calendar to help you answer these questions.

1. How many in a year?

days **weeks**

months

2. If January is the **first** month, what month is

March

June

September

3. Sometimes dates are written with numbers.
Here's an example:
The third day of July, 1994 can also be written: **3/7/94** (day, month, year)

How would you write these dates in numbers?

The fifteenth day of April, 1991

The twentieth day of August, 1989

What is today's date in numbers?

4. What number do you add to find out what the date will be tomorrow?

5. What number will you add to find the date one week from now?

6. What number do you subtract to find the date one week ago?

7. How many days are in

5 weeks? 7 weeks?

8. If today is the 25th, what was the date 6 days ago?

If there are only 30 days in the month, what will the date be in 6 days?

9. If there are 30 days in a month, how many full weeks are in it?

10. Take today's date and write as many equations as you can with that number as the answer. Try to do a variety of operations. *For example*, if today is the 15th:

3x5 13+2 28-13 30-:-2 2+2+6+2+2+1 (2x4)+10}-3

REAL WORLD MATH

Schedules

Your Experience

You see your parents write appointments in a planning book. They will refer to it often thoughout the day.

Let's Explain It

People have found that planning their day increases their productivity. They must record all appointments to schedule around so that they remember them.

Practice Page Directions

1. Choose from the list of activities to help you plan a day in the appointment book. You will need to allow enough time for each as you schedule it into the day.
2. Prioritize the things you have to do by listing them in importance, first (1st) through (5th), with "A" being the most important.

MONDAY JANUARY 18

8:00	Go to School
9:00	
10:00	
11:00	Lunch
12:00	
1:00	Dentist
2:00	Soccer Practice
3:00	
4:00	
5:00	

TUE
8:00
9:00
10:00
11:00
12:00
1:00
2:00
3:0
4:

Real World Activity

- Ask some adults to share their daily appointment planner with you. What type of planner do they use and why? Have they taken a course in time management?
- Buy an inexpensive appointment book. Keep track of your plans and assignments. Does it help you to be more organized?

Things to Think About

Why is planning necessary?

Why is prioritizing important?

When should a daily planner be kept?

How is an assignment notebook similar to a planner?

Words to Know

- schedule
- appointment
- ordinal number
- agenda
- organize
- prioritize

Schedules

Practice Page Directions

1. Choose at least six activities from the list and plan a day in the appointment book. Be sure to allow enough time for each activity as you schedule it into the day.
2. Prioritize the things you have to do by listing them by importance, first (1st) through sixth (6th) or as many as you need. Do this just in case you don't have enough time for everything!

Homework	*1 hour*
Household chores	*30 minutes*
Library	*1 hour*
Get ready for school	*1 hour*
School	*6 hours*
Club meeting	*1 hour*
Plan surprise party	*45 minutes*
Read new book	*30 minutes*
Sports practice	*1 hour 30 minutes*
Dance class	*1 hour*
Shopping	*1 hour 15 minutes*
Dinner	*30 minutes*
Telephone calls	*15 minutes*
Paper route	*45 minutes*
Get ready for bed	*15 minutes*

Appointment Schedule

Date:

Time	Activity	Priority

REAL WORLD MATH

Carpenter Calculations

Your Experience

Your family is building a fence for your dog. Your dad figured the amount of wood fencing needed to build the fence.

Let's Explain It

The distance around an area is called the *perimeter*. By adding the length of all four sides, the amount of fence needed can be determined.

Practice Page Directions

1. Look at the dogs' yards on the next page.
2. Add all four sides together to calculate the distance around the yard.
3. For a short cut, use multiplication for sides of the same length.

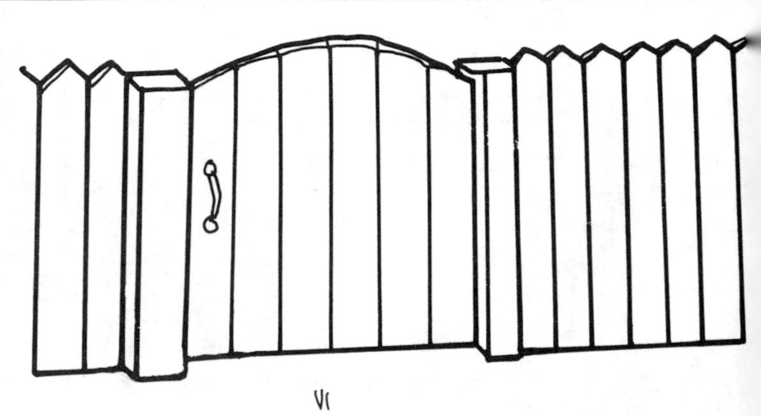

Real World Activity

- Measure the length of the sides of your backyard.
- Calculate the perimeter.
- Draw a diagram of it. How much fence will you need to enclose your yard? Have your parents check your work.

Things to Think About

What other math skills would a carpenter use in his or her work?

Why is measurement important for building?

What other jobs might require calculating the perimeter?

Words to Know

- perimeter
- yards
- length
- area
- meters
- width

Carpenter Calculations

Practice Page Directions

1. Here are three dogs who need fences around their yards. Help the carpenter on the job by figuring out the perimeter measurement in standard or metric.
2. Add all four sides together to calculate the perimeter, the distance around the yard.
3. For a short cut, use multiplication for sides of the same length.

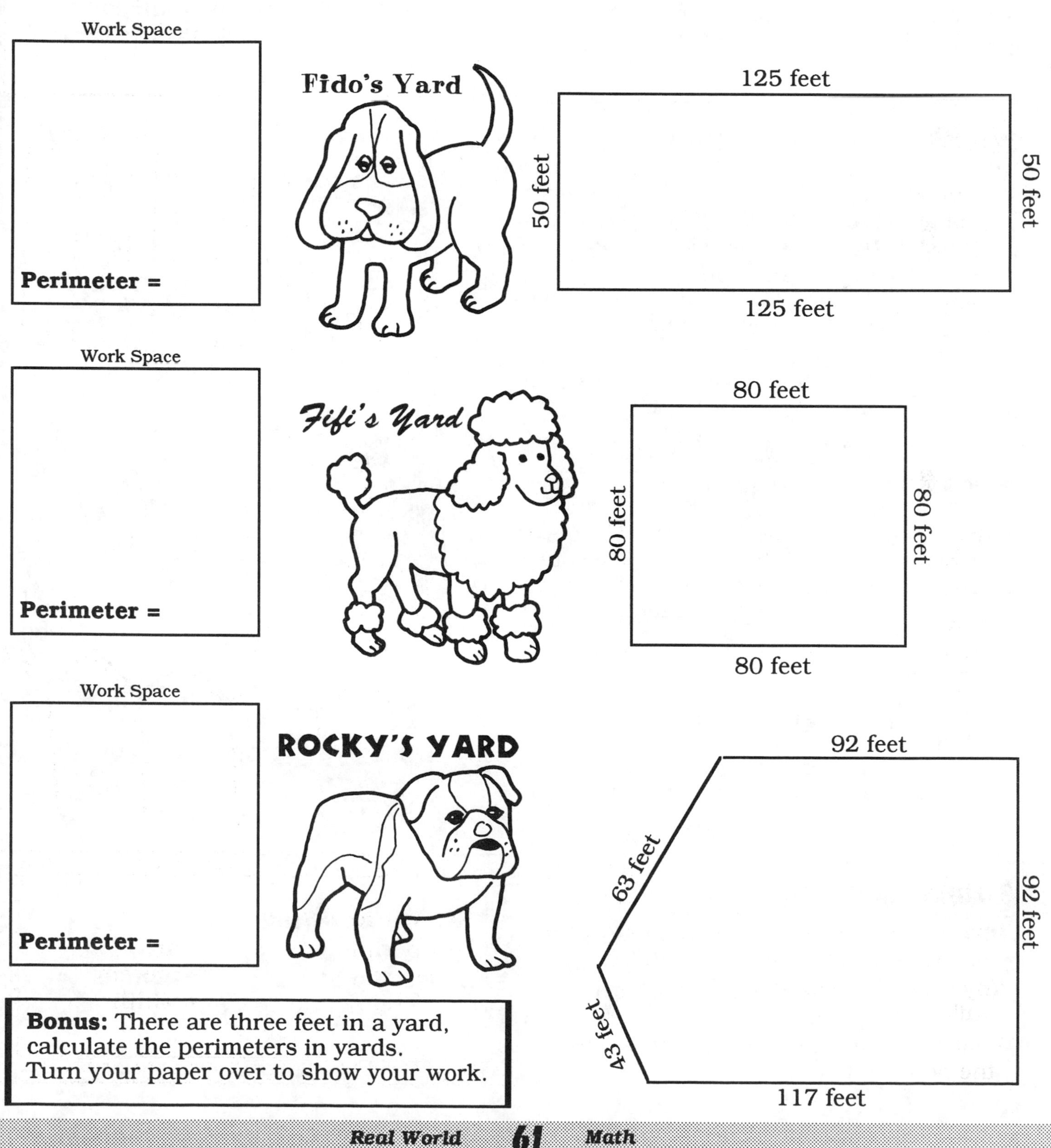

Bonus: There are three feet in a yard, calculate the perimeters in yards.
Turn your paper over to show your work.

R E A L W O R L D MATH

Grocery Shopping

Your Experience

Your trip to the grocery store was a disaster! When you checked out you didn't have enough money to pay for the food you bought.

Let's Explain It

Check the grocery store ads for items on sale and make a list of things to buy before you go to the store so that you don't spend more than you have.

Practice Page Directions

1. Look at the grocery store shelves. Make a shopping list of the things you plan on purchasing. Write their cost.
2. Estimate how much money you think you will spend. Round the price of the items to help with your estimation.
3. Add the total price. Use a calculator to check your addition. How close was your estimation to the actual price?

Real World Activity

- Use newspaper advertisements to make a shopping list of ten items. Estimate your total cost. With the permission of your parents, and perhaps their money, go to the store with that amount of money and purchase the ten items. Did you estimate and budget correctly?
- Go on a grocery store shopping trip with an adult. Take a calculator. Total each item as it is put in the shopping cart. How close was your total to the actual amount spent?

Things to Think About

How much money does your family spend on groceries each week? month?

What could you do if the amount of money you brought with you to the store was less than the cost of the items you actually purchased?

Words to Know

- estimate
- calculate
- comparison shopping
- coupons
- budget

Practice Page Directions

Grocery Shopping

1. Look at the items on the grocery store shelves. Make a shopping list of the things you plan on purchasing. Include their cost in the column on the right side of the list.
2. Estimate how much money you think you will spend. Round the price of the items up to help with your estimation. Write your estimate in the box at the bottom of the page.
3. Add the total price of the tickets using this estimated amount. Use a calculator to check your addition. Determine the exact cost without tax. How close were you to the actual price?

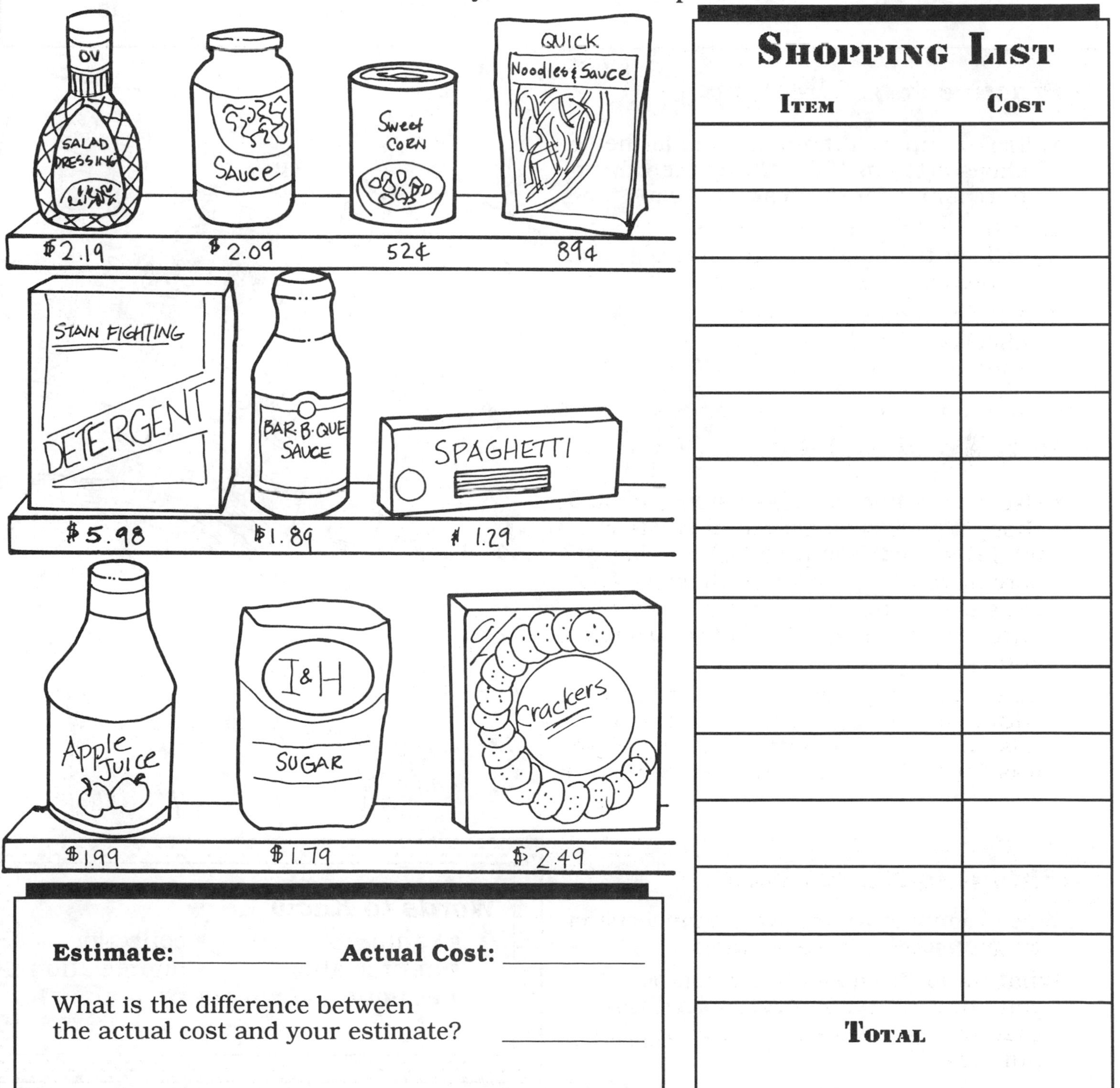

Shopping List

Item	Cost
Total	

Estimate:__________ **Actual Cost:** __________

What is the difference between the actual cost and your estimate? __________

REAL WORLD MATH

Collections

Your Experience

Your friends all have collections such as baseball cards, stamps and coins. You want to start a collection. Which one would be the most profitable?

Let's Explain It

Most people do not start collections for their value, but rather because they are interested in them. Most collections will increase in value as time passes.

Practice Page Directions

1. The stamp collection pictured in the box was started in 1985. Study the value of the stamps for the years following the start of the collection.
2. Answer the questions about the collection.

Real World Activity

- Go to the library or a bookstore to find out the real prices of a collection in which you are interested. There are many collector's books available in which values of collectibles are listed.
- Check out how much some of your collections are worth.

Things to Think About

What factors make an item valuable other than its age?

What things from the 90 s might be valuable in 20 years? A hundred?

Where would you go to sell your collection?

Words to Know

- value
- appreciation
- antique
- collection
- depreciation
- profitable

Practice Page Directions

1. The stamp collection below was started in 1985. Study the value of the stamps for the years following the start of the collection.
2. Answer the questions about the collection.

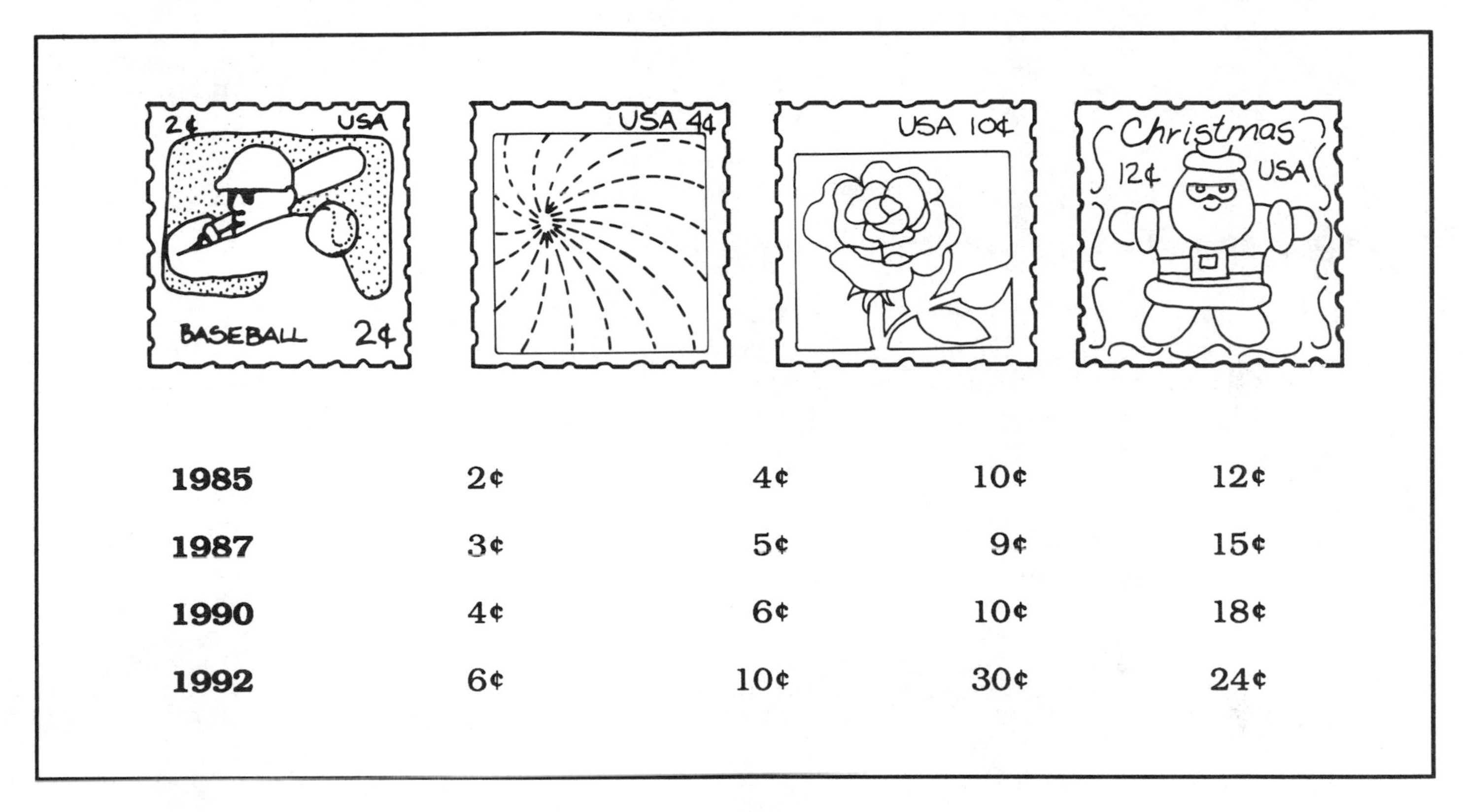

1985	2¢	4¢	10¢	12¢
1987	3¢	5¢	9¢	15¢
1990	4¢	6¢	10¢	18¢
1992	6¢	10¢	30¢	24¢

1. Circle the stamp that gained the most value.

2. How much was the whole collection worth in 1985? __________ in 1992?__________

3. What was the percentage of increase over the seven years from when the collection started through 1992? (Total value divided by amount of increase)

4. Put an X through the stamp that earned the least.

5. In which year did the value of the whole collection increase the most?

REAL WORLD MATH

Dining Out

Your Experience

Your family goes out to dinner once a week. You want to make it twice a week. Can your family afford do to that?

Let's Explain It

Some restaurants are more expensive than others. If you plan and calculate correctly, your savings may pay for an additional time to eat out.

Practice Page Directions

1. Look at the two menus on the next page. Calculate the cost of your meal at each restaurant. Multiply it by four or the number of members in your family.
2. What is your savings at the cheaper restaurant?
3. Would the savings pay for another evening of eating there?

Real World Activity

- The next few times your family eats out, record what was eaten and how much it cost at each different restaurant. What was the difference in cost between the two places? Keep a record in your *Real World Journal.*

Things to Think About

How much would it cost your family to cook that same meal at home?

What is the percentage of increase to eat that meal at a restaurant?

How much should you tip at a sit down restaurant?

Words to Know

- restaurant
- tip
- entrée
- change
- percentage
- expensive

Practice Page Directions

Dining Out

1. Look at the two menus below.
2. "Order" a meal from each by circling the items you chose. Calculate the cost of your meal at each restaurant. Multiply it by four **or** the number of members in your family.
3. Circle the meal that is less expensive. Figure out the difference in price.

Rose's Cafe

Hamburger with fries, cole slaw $5.95

Add cheese .50¢

Chef Salad $5.50

Hot dog basket $3.25

Coke $1.25

Lemonade $1.00

McBurger's Drive-Thru

Hamburger $1.50

Cheeseburger $1.75

Salad $2.25

Fries $1.00

Coke .75¢, .85¢, .99¢

Hot dog $1.00

Circle the food you would select off of the menu and its cost. In the space below each restaurant name, figure out how much you spent on one meal.

Rose's Cafe

Total=

McBurger's Drive-Thru

Total=

Work Space

1. Multiply each total times the number of people in your family.
2. Calculate the tip for service at Rose's Cafe. A standard tip is approximately 15% of the food bill. ___ X 15%=___
3. Add this on to the cost of the meals.
4. What is the difference in price between the two restaurants?
5. If you handed the waiter a $100 bill, how much change would you get back?

1.

2.

3.

4.

5.

REAL WORLD MATH

Graphs

Your Experience

Your teacher makes you read graphs at school because she says they are useful ways of showing information. How do they do this?

Let's Explain It

A graph can show numerical information in a visual way. There are many kinds of graphs that can be used to show the same statistics.

Practice Page Directions

1. Study the two different graphs. Answer the questions about each.

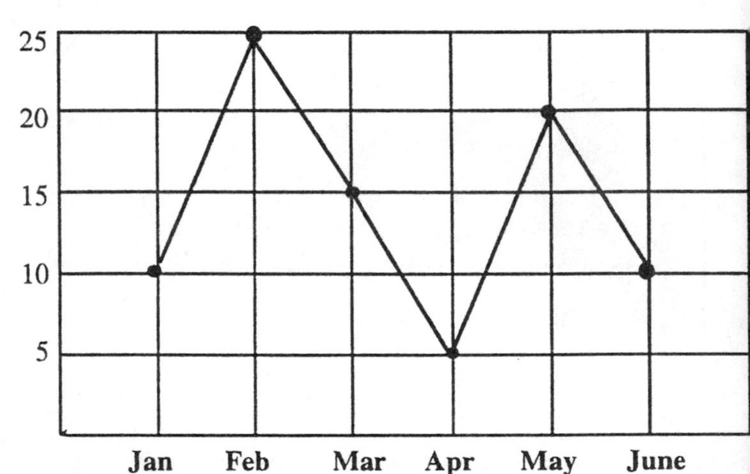

Real World Activity

- Go on a scavenger hunt with a friend to see who can find and cut out the most graphs in newspapers or magazines. Set a time limit. Count how many are found.
- Show each other the graphs and try to interpret them. Give titles to each graph.

Things to Think About

How many different kinds of graphs are there?

How are graphs used in business? What business would use graphs the most?

How are graphs used to show survey results?

Words to Know

- pie/bar/line graph
- vertical/horizontal
- interpret
- data
- percent
- line graph

Study the two different graphs. Answer the questions about each.

Graphs

PIE SURVEY

50 people were asked to name their favorite pie. The results are shown here in—what else?—a ***pie graph****.*

1. What percentage liked apple the most?

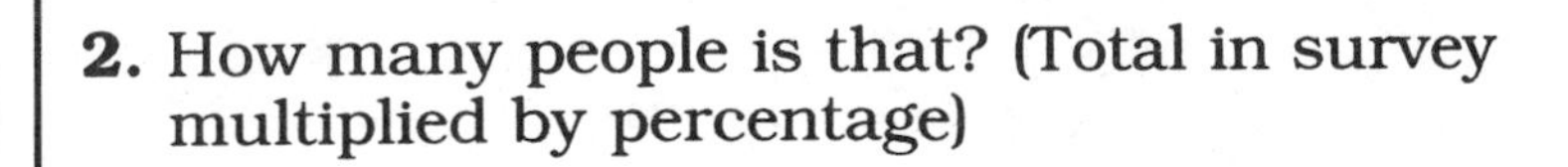

2. How many people is that? (Total in survey multiplied by percentage)

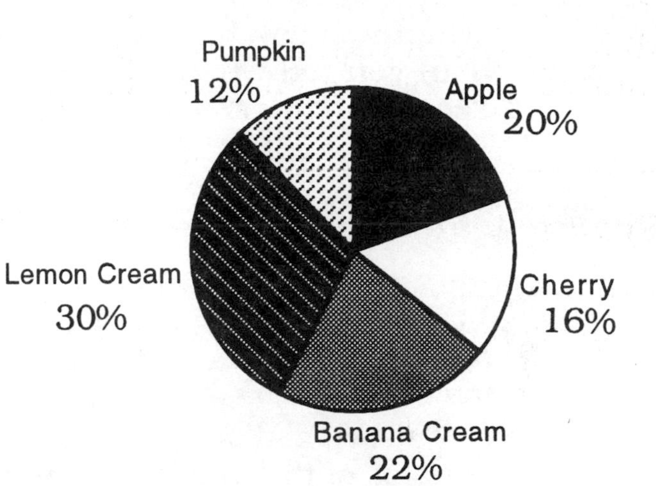

3. Which kind was preferred the least?

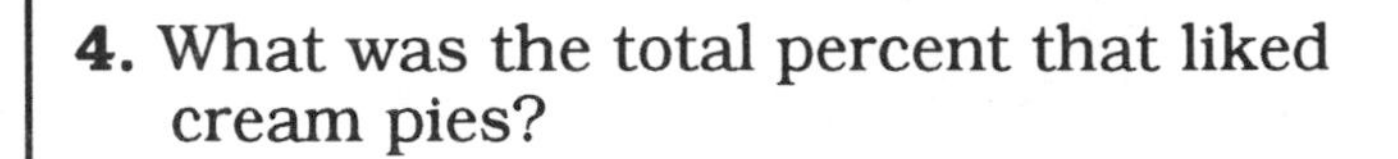

4. What was the total percent that liked cream pies?

5. Write the name of the pies in order of preference from most-liked to least-liked.

ANOTHER PIE SURVEY!

Molly's Pie Shop was asked to graph sales of their famous cherry pie for the first six months of the year. The results are shown in the line graph/grid below. Interpret the data.

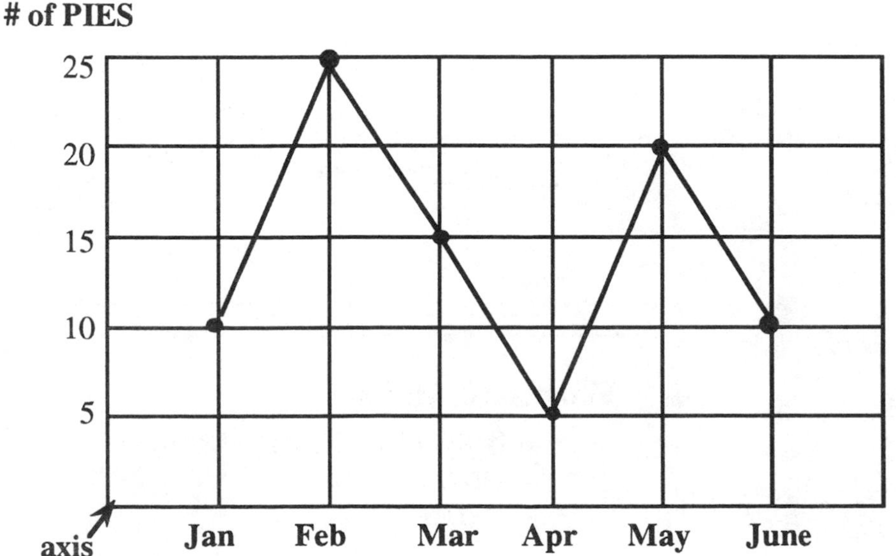

1. During which month were the most cherry pies sold?

2. Were there any months that had equal sales?

3. How many pies were sold during the sixth month period?

4. If you owned Molly's Pie Shop, which month would you offer a special low price on cherry pie? Why?

REAL WORLD MATH

Let's Go Fishing

Your Experience

You love to go fishing with your family. It's a lot more fun, however, when you catch a fish or two. How can you tell when you've caught a record breaker?

Let's Explain It

A fish can be measured for its length and weighed for its size. A spring scale can be purchased in the sporting goods department to weigh your fish.

Practice Page Directions

1. Look at the fish caught in the fishing contest! Use the chart to help figure the total length of the fish caught by each fisherman (or woman). You'll need to multiply and add!
2. Write the name of the winner—the one with the highest total length—in the ribbon!

Real World Activity

- Keep a record in your *Real World Journal* of the fish you catch in the next month. Plot the length and weight on a graph. Ask an adult fisherman to record his fish caught over a month. Compare the record with yours.
- If you aren't a fisherman, look in the fishing section of the newspaper. What is the largest fish caught by a professional this week?

Things to Think About

Could a fish be shorter than another and still weigh more?

What is the largest fish caught according to the *Guinness Book of World Records?*

Are professional fisherman paid for what they catch?

Words to Know

- scale
- length
- weight
- measure
- average
- record

Let's Go Fishing

Practice Page Directions

1. Look at the fish caught in the fishing contest! Use the chart to help figure the total length of the fish caught by each fisherman (or woman). You'll need to multiply and add!
2. Write the name of the winner—the one with the highest total length—in the ribbon!

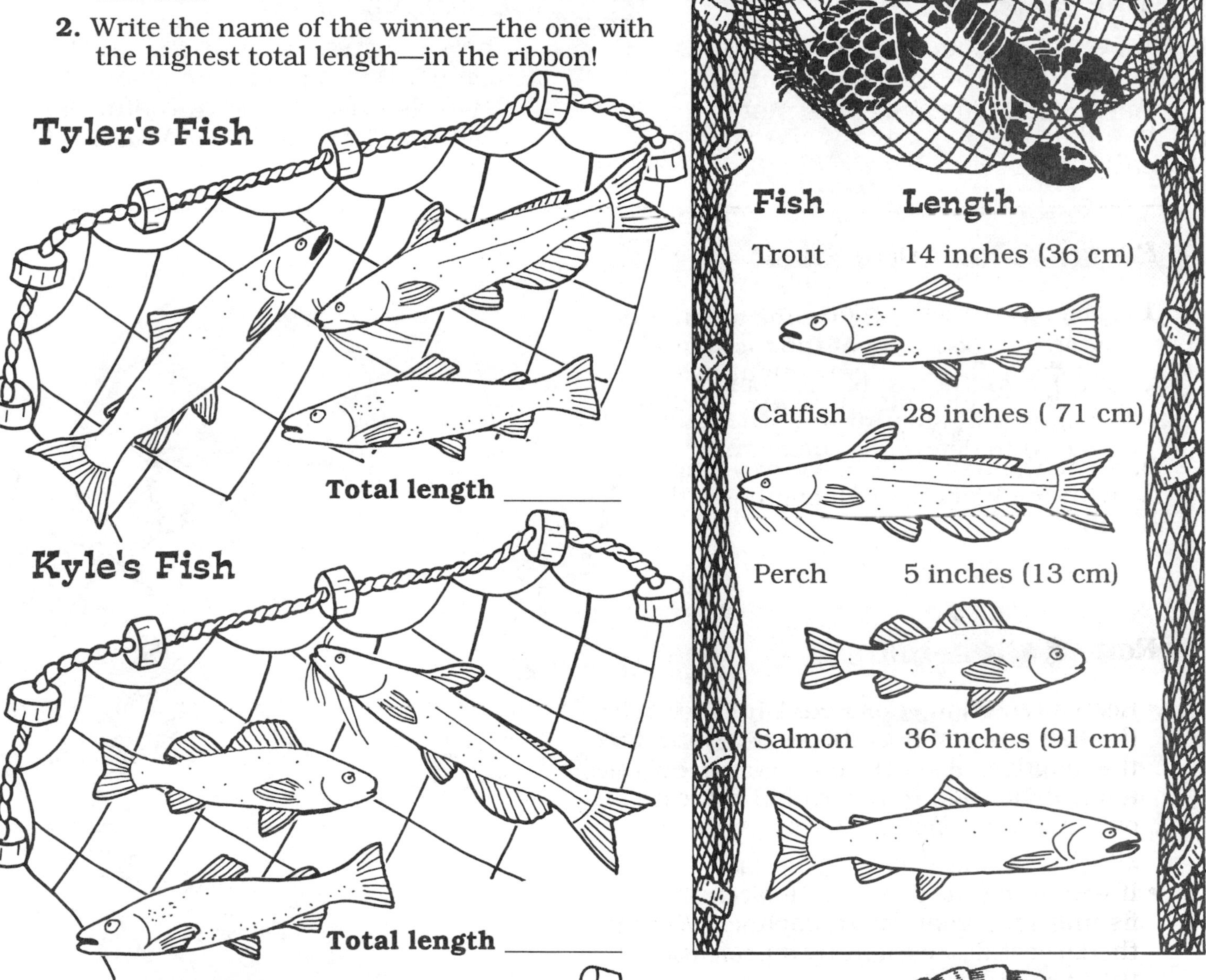

Fish	Length
Trout	14 inches (36 cm)
Catfish	28 inches (71 cm)
Perch	5 inches (13 cm)
Salmon	36 inches (91 cm)

Tyler's Fish

Total length ________

Kyle's Fish

Total length ________

Terry's Fish

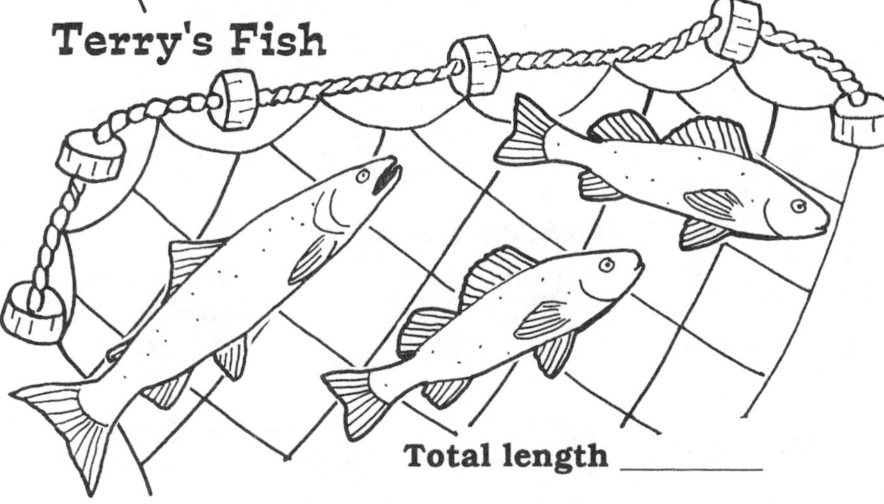

Total length ________

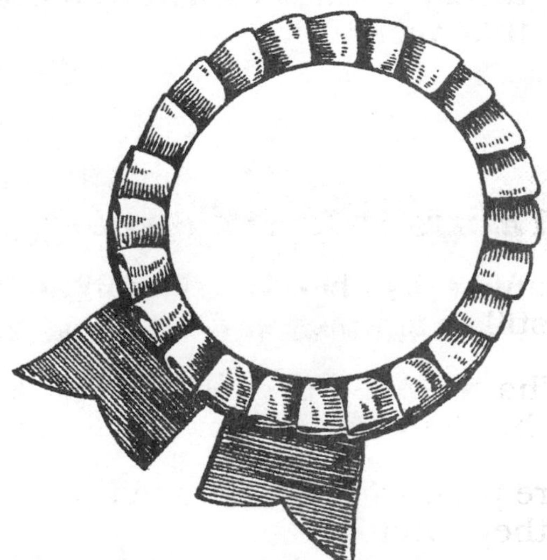

REAL WORLD MATH

Sales Tax

Your Experience

You see the price tag on items at the store. However, when you check out, it always costs more. Why is this?

Let's Explain It

The store must charge tax on the sale of all items. The amount depends on the local and state tax level that has been voted upon by the public.

Practice Page Directions

1. Use a calculator to figure the amount of tax charged on each of the items.
2. Add this to the cost of the item.
3. Subtract that amount from the total received to determine the change.

$3.50

$8.75

37.45

Real World Activity

- Take a list of items your family needs to purchase from the store. Approximate the total cost and calculate sales tax according to your city. Add it to the total amount.
- After the items are purchased, see how close you were. You can also take some sales receipts and check the accuracy of the sales tax charged on each.

Things to Think About

How is the sales tax you pay used?

Do you know what percentage goes where?

Why are different cities charging different sales taxes?

Words to Know

- sales tax
- receipt
- additional
- rate
- taxable item
- percent

Practice Page Directions

Sales Tax

1. Use a calculator to figure the amount of tax charged on each of the items.
2. Add this to the cost of the item.
3. Subtract that amount from the total received to determine the change.

1. Item $5.00

Sales tax 10%

Tax= $________

Total Price

Amount given $10.00

Change received $____________

2. Item $2.00

Sales Tax 6.5%

Tax= $________

Total Price

Amount given $5.00

Change received $____________

3. Item $3.50

Sales Tax 8%

Tax= $________

Total Price

Amount given $5.00

Change received $____________

4. Item $8.75

Sales Tax 7%

Tax=$________

Total Price

Amount given $10.00

Change received $____________

5. Item $12.00

Sales Tax 5.5%

Tax=$________

Total Price

Amount given $20.00

Change received $____________

6. Item $22.00

Sales Tax 5.5%

Tax=$________

Total Price

Amount given $25.00

Change received $____________

REAL WORLD MATH

Cooking

Your Experience

Your mom fixes your favorite food, macaroni and cheese, for lunch. She says it is simple enough for you to make. The directions are right on the box.

Let's Explain It

Reading and following directions, measuring and timing are some of the math skills involved in cooking. Accuracy is important for the final product to be tasty.

Practice Page Directions

1. Read the directions for fixing macaroni and cheese. Answer the questions beside the box to show you understand.
2. Double the recipe at the bottom of the page.
3. Choose two extra ingredients from the choices listed.

Real World Activity

- Fix a box of macaroni and cheese for your entire family at dinner. Add other foods to balance your meal. Prepare them by reading the directions on the box. Set the dinner table. How is math used? Invite your family back into the kitchen to enjoy the food.

Safety tip: Be sure you know how to use the stove or microwave before preparing the meal on your own.

Things to Think About

What is the difference between liquid and solid measurement?

How are recipes created?

Why is accurate measurement important in preparing a recipe?

Think of ways to change a favorite recipe without affecting the base recipe.

Words to Know

- liquid and dry
- ounces/cups
- temperature
- quart/liter
- degrees
- yield
- teaspoon
- tablespoon

Practice Page Directions

Cooking

1. Read the directions for fixing macaroni and cheese. Answer the questions below the box to show you understand.
2. Double the recipe at the bottom of the page.

Macaroni & Cheese

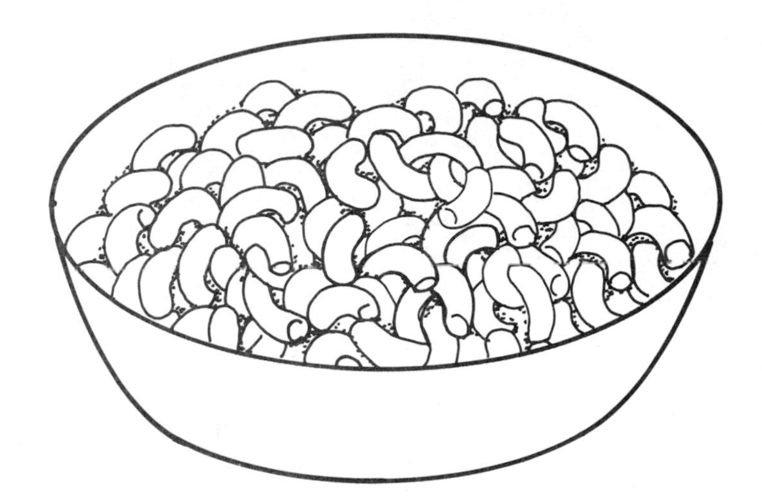

Stove Top Directions

1. Add 2 cups macaroni and 1 teaspoon salt to 6 cups boiling water in 4-quart pan.
2. Boil rapidly, stirring occasionally, 5 to 6 minutes or to desired tenderness.
3. Drain. Add ¼ cup milk, ¼ cup (½ stick) butter and 2 teaspoons seasoning.
4. Heat 3 minutes on low. Serve.

Yield: Four ⅔ cup servings

Optional: Add one of the following items:

1 cup chopped cooked ham

¼ cup salsa

16 oz. can vegetables, drained

1 cup chopped cooked chicken or turkey

2 tablespoons bacon bits

1. If there are 4 cups of water in a quart, how many cups would you need to fill the pan?
2. What is the total cooking time?
3. How many total cups of macaroni and cheese will this recipe yield?
4. List the liquid measurements you will make:
5. List the dry measurements you will make:

- Calculate how much of each you would need if you double the recipe:

___ macaroni

___ teaspoons of salt

___ cups boiling water

___ cups milk

___ cups butter

___ seasoning

___ optional ingredient ___________

Yield: ____ ⅔ cup servings

REAL WORLD MATH

Catalog Shopping

Your Experience

Your family gets a lot of catalogs in the mail. You see something you want! How do you know how much money to send?

Let's Explain It

Every catalog has an order form that has places for you to write what you want, what size and the total cost. You need to calculate carefully to send the right amount.

Practice Page Directions

1. Locate several catalogs around the house. Your parents can tell you where they are.
2. Look through them and select four things you would like to buy. Cut them out and tape them in the box.
3. Complete the order form. Calculate all of the costs!

Real World Activity

- When your parents need to order an item from a catalog, ask them if you can do it for them. Complete the order blank. Have your parents check your calculations. Include the order form with the check or money order. If you are ordering by phone, have the order form ready.
- Count the number of days it takes to receive your order. Is everything correct? Did you need to send any more money?

Things to Think About

How is your order filled once it is received at the factory?

What happens if you make a calculation mistake on the order form?

How much more do you pay for something when you order it by mail?

Words to Know

- quantity
- handling charge
- shipping charge
- sales tax
- money order
- total

Practice Page Directions

Catalog Shopping

1. Locate several catalogs around the house. Your parents can tell you where they are.
2. Look through them and select four things you would like to buy. Cut them out and tape them to the box below.
3. Complete the order form. Calculate all the costs!

Things I want to order:

Home Shopping Order Form

Item #	Description	Size	Color	Quantity	Price	Total Price

Shipping Charges

Orders Total	Add
Up to $20.	$4.00
$20.01 to $30.	$5.00
$30.01 to $40.	$6.00
$40.01 to $50.	$7.00
Over $50.	$8.00

For faster service, we can arrange for Express Shipping (3 to 4 days). The additional charge will be $5.00.

Total Merchandise	
Sales Tax 5%	
Shipping (See box at left)	
Express Ship (See info at left)	
Total Amount Enclosed	

Answer Key

Credit Cards Pages 8-9
See student work

Classified Advertisements Pages 10-11
See student work

Checkbook Pages 12-15
See student work

Buying Produce Pages 16-17
Peaches—$3.95
Oranges—$4.87
Zucchini—$2.52
Broccoli—$3.06

Sports Scores Pages 18-19
Football
Team A—27
Team B—26
Winning Team—Team A

Basketball
Team A—96
Team B—101
Winning Team—Team B

Sports Stastics Pages 20-21
Batting Averages

Brett	(.166)
Otis	(.100)
Arnold	(.500)
Samantha	(.320)
Frank	(.142)
Jose	(.125)
Thomas	(.225)
Pat	(.107)
Martin	(.240)

Batting Order
1. Arnold
2. Samantha
3. Martin
4. Thomas
5. Brett
6. Frank
7. Jose
8. Pat
9. Otis

Awards
Home Run King: Thomas
Top Hitter: Martin
Least Errors: Frank
Most Valuable: see student work

Calories & Grams Pages 22-23
1. 160 calories, 10 calories from fat
2. 60 grams of fat
3. No standard answer
4. 598 calories, 31.8 grams fat
5. No standard answer
6. 5 fat grams
7. 99¢ per glass, $9.99
8. 29.5 grams

Making Change Pages 24-25
Baseball, mitt, dice, ice cream:
Total—$3.70
Change from $10.00—$6.30
Change from $20.00—$16.30

Baseball, marbles, doll, teddy bear:
Total—$4.29
Change from $10.00—$5.71
Change from $20.00—$15.71

Doll, bunny, marbles, cupcake:
Total—$4.75
Change from $10.00—$5.25
Change from $20.00—$15.25

Teddy bear, doll, marbles, baseball:
Total—$4.29
Change from $10.00—$5.71
Change from $20.00—$15.71

Food Fractions Pages 26-27
See student work

Coupons Pages 28-31
Page 29
1. $2.30
2. 29¢
3. Two
4. 50¢
5. Cereal Raisin
6. $1.25

Pages 30-31
See student work

Budgets Pages 32-33
See student work

Allowance Pages 34-35
Bike—20 weeks, 12¼ weeks, 28 weeks, 39¼ weeks, 14 weeks
Doll—3½ weeks, 2¼ weeks, 5¼ weeks, 7¼ weeks, 2½ weeks
Ticket—5¼ weeks, 3¼ weeks, 7½ weeks, 10½ weeks, 3¾ weeks
Skateboard—8½ weeks, 5¼ weeks, 12 weeks, 16¾ weeks, 6 weeks
Book—2 weeks, 1¼ weeks, 2¾ weeks, 4 weeks, 1½ weeks

Postage Stamps Pages 36-37
See student work
Postage needed:
Carrie Becker—29¢
Jason Smith—19¢
Maria Lopez—50¢
Anita Parker—98¢
Leanne Milliken—52¢

Movie Outing Pages 38-39
Matinee, nachos, drinks—$28.00
Evening, popcorn, three drinks, licorice, candy—$29.25
Evening, four popcorns, four drinks, four gum, four candy—$41.00
Matinee, two licorice, two candy, four drinks—$21.00
See student work

Income Pages 40-43
Electrical wiring—$7.00 hourly, $14.00 total
Consulting—$10.00 hourly, $30.00 total
Painting—$8.00 hourly, $96.00 total
Trash removal—no hourly rate, $6.00 total
Trash Bags—no hourly rate, $2.00 total
Invoice Total—$148.00

Pages 42-43
See student work

Sale Prices Pages 44-45
Answers are listed in this order: Original Price, Sale Price, Savings
Shirt—$16.00, $12.00, $4.00
Dress—$25.00, $18.75, $6.25
Coat—$65.00, $48.75, $6.25
Tea Kettle—$24.00, $12.00, $12.00
Clock—$9.99, $5.00, $4.99
Bag—$34.00, $17.00, $17.00
Candy Dish—$12.99, $6.50, $6.49
Glove—$13.00, $6.50, $6.50
Earrings—$5.49, $4.94, 55¢
Cologne—$12.99, $11.69, $1.30
Glasses—$8.75, $7.87, 88¢
Slippers—$14.50, $13.05, $1.45

Fund-raising Pages 46-47
Answers are listed in this order: Gross Profit, New Profit School, Net Profit Company
Gift Wrap—$281.25, $156.25, $125.00
Candy Bars—$217.00, $75.95, $141.05
Calendars—$318.00, $153.70, $164.30
Total: $385.90
See student work

Clocks Pages 48-49
See student work

Automobile Gauges Pages 50-51
1. B
2. C
3. B
4. B
5. A
6. 115,808
7. 25 mph
8. 23

Buying a Car Pages 52-53
$7,500 car—$229.16 payment, $11.46 penalty
$12,500 car—$286.45 payment, $14.32 penalty
$17,350 car—$312.30 payment, $15.62 penalty
$25,070 car—$417.83 payment, $20.89 penalty

Calculators Pages 54-55
25 x 42 = 1050
160 ÷ 4 = 40
2580 + 3719 = 6299
8691 - 3176 = 5515
2 x 18 x 4 + 200 = 344
(477 ÷ 30) - 12 = 3.9
679 - 214 + 75 ÷ 6 = 90

90 + (13 x 4) = 142
5948 - 4679 = 1269
(150 ÷ 3) x 4 = 200

Calendar Pages 56-57

1. Seven days, 52 weeks, 12 months
2. March—3rd month, June—6th month, September—9th month
3. 4/15/91, 8/20/89, see student work
4. 1
5. 7
6. 7
7. 35 weeks, 49 weeks
8. 19th, see student work
9. 4
10. See student work

Schedules 58-59

See student work

Carpenter Calculations Pages 60-61

Fido— 350 feet; bonus answer—116 yards, 2 feet
Fifi—320 feet; bonus answer—106 yards, 2 feet
Rocky—407 feet; bonus answer—135 yards, 2 feet

Grocery Shopping 62-63

See student work

Collections Pages 64-65

1. 10¢
2. 28¢, 70¢
3. 166%
4. 2¢
5. 1992

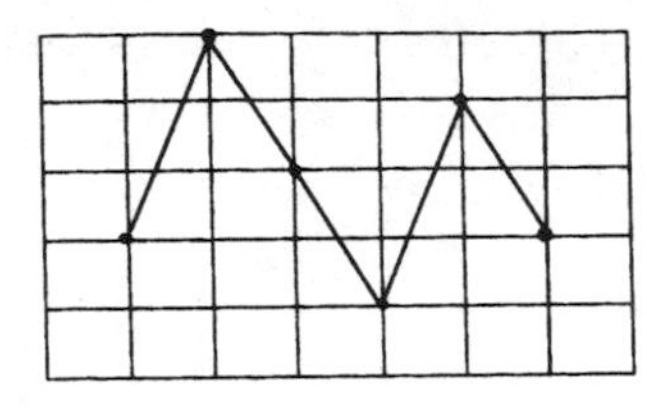

Dining Out Pages 66-67

See student work

Graphs Pages 68-69

1. 20%
2. 10
3. pumpkin
4. 55%
5. Lemon cream, banana, apple, cherry, pumpkin

Let's Go Fishing Pages 70-71

Tyler's Fish—78 inches/198 cm
Kyle's Fish—47 inches/120 cm
Terry's Fish—46 inches/117 cm

Sales Tax Pages 72-73

Answers are listed in the following order: Amount of Tax, Total Price, Change Received

1. 50¢, $5.50, $4.50
2. 13¢, $2.13, $2.87
3. 28¢, $3.78, $1.22
4. 61¢, $9.36, 64¢
5. 66¢, $12.66, $7.34
6. $1.21, $23.21, $1.79

Cooking Pages 74-75

1. 16 cups
2. Eight to nine minutes
3. 2⅔ cups
4. 6 cups of water, ¼ cup of milk
5. 2 cups of macaroni, 1 tsp. salt, ¼ cup of butter

• 4 cups of macaroni, 2 tsp. salt, 12 cups of water, ½ cup of milk, ½ cup butter, 4 tsp. seasoning; optional ingredient varies; yield 8 servings

Catalog Shopping 76-77

See student work